# EDUCATION, POLITICS, AND PUBLIC LIFE

Series Editors:
Henry A. Giroux, McMaster University
Susan Searls Giroux, McMaster University

Within the last three decades, education as a political, moral, and ideological practice has become central to rethinking not only the role of public and higher education, but also the emergence of pedagogical sites outside of the schools—which include but are not limited to the Internet, television, fi lm, magazines, and the media of print culture. Education as both a form of schooling and public pedagogy reaches into every aspect of political, economic, and social life. What is particularly important in this highly interdisciplinary and politically nuanced view of education are a number of issues that now connect learning to social change, the operations of democratic public life, and the formation of critically engaged individual and social agents. At the center of this series will be questions regarding what young people, adults, academics, artists, and cultural workers need to know to be able to live in an inclusive and just democracy and what it would mean to develop institutional capacities to reintroduce politics and public commitment into everyday life. Books in this series aim to play a vital role in rethinking the entire project of the related themes of politics, democratic struggles, and critical education within the global public sphere.

## SERIES EDITORS:

**HENRY A. GIROUX** holds the Global TV Network Chair in English and Cultural Studies at McMaster University in Canada. He is on the editorial and advisory boards of numerous national and international scholarly journals. Professor Giroux was selected as a Kappa Delta Pi Laureate in 1998 and was the recipient of a Getty Research Institute Visiting Scholar Award in 1999. He was the recipient of the Hooker Distinguished Professor Award for 2001. He received an Honorary Doctorate of Letters from Memorial University of Newfoundland in 2005. His most recent books include *Take Back Higher Education* (co-authored with Susan Searls Giroux, 2006); *America on the Edge* (2006); *Beyond the Spectacle of Terrorism* (2006), *Stormy Weather: Katrina and the Politics of Disposability* (2006), *The University in Chains: Confronting the Military-Industrial-Academic Complex* (2007), and *Against the Terror of Neoliberalism: Politics Beyond the Age of Greed* (2008).

**SUSAN SEARLS GIROUX** is Associate Professor of English and Cultural Studies at McMaster University. Her most recent books include *The Theory Toolbox* (co-authored with Jeff Nealon, 2004) and *Take Back Higher Education* (co-authored with Henry A. Giroux, 2006). Professor Giroux is also the Managing Editor of *The Review of Education, Pedagogy, and Cultural Studies.*

*Critical Pedagogy in Uncertain Times: Hope and Possibilities*
Edited by Sheila L. Macrine

*The Gift of Education: Public Education and Venture Philanthropy*
Kenneth J. Saltman

*Feminist Theory in Pursuit of the Public: Politics, Education, and Public Life*
Robin Goodman (forthcoming)

*Hollywood's Exploited: Public Pedagogy, Corporate Movies, and Cultural Crisis*
Edited by Anthony J. Nocella, II, Rich Van Heertum, Benjamin Frymer, and Tony Kashani (forthcoming)

*Education out of Bounds: Cultural Studies for a Posthuman Age*
Richard Kahn and Tyson Lewis (forthcoming)

*Rituals and Student Identity in Education: Ritual Critique for a New Pedagogy*
Richard A. Quantz (forthcoming)

# The Gift of Education

## Public Education and Venture Philanthropy

*Kenneth J. Saltman*

First published in 2010 by
PALGRAVE MACMILLAN®
in the United States—a division of St. Martin's Press LLC,
175 Fifth Avenue, New York, NY 10010.

Where this book is distributed in the UK, Europe and the rest of the world,
this is by Palgrave Macmillan, a division of Macmillan Publishers Limited,
registered in England, company number 785998, of Houndmills,
Basingstoke, Hampshire RG21 6XS.

Palgrave Macmillan is the global academic imprint of the above companies
and has companies and representatives throughout the world.

Palgrave® and Macmillan® are registered trademarks in the United States,
the United Kingdom, Europe and other countries.

ISBN 978–0–230–61514–4 Hardcover
ISBN 978–0–230–61515–1 Paperback

Library of Congress Cataloging-in-Publication Data

Saltman, Kenneth J., 1969–
    The gift of education: public education and venture philanthropy /
    Kenneth J. Saltman.
        p. cm.— (Education, politics, and public life)
    Includes bibliographical references and index.
    ISBN 978–0–230–61514–4 (alk. paper)
        1. Public schools—United States. 2. Education—Charitable
    contributions—United States. I. Title.

LA217.2.S256 2010
379.1—dc22                                              2009039974

Design by Newgen Imaging Systems (P) Ltd., Chennai, India.

First edition: March 2010

10 9 8 7 6 5 4 3 2 1

Printed in the United States of America.

*For Kathy*

# Contents

# LIST OF TABLES

# Acknowledgments

This book began about a decade ago with the reading I did on general economy, particularly books by Georges Bataille and Jean Baudrillard, the discussions of which ended up getting cut from the final manuscript. After writing my first book *Collateral Damage* about the corporatization of schools, I thought that it would be worthwhile to expand on a line in that book about how teachers are motivated to teach by a desire to give selflessly to others with no expectation of return. In the years since writing that, the mistaken assumption that the "right knowledge" must be enforced through disciplinary measures has only expanded. This book took its current form more recently when that early interest in thinking about education beyond economism became concretized through focusing it on the rise of venture philanthropy. The more I looked at right-wing think tanks and their role in the privatization and deregulation agendas for my book *Capitalizing on Disaster*, the more I found that the new philanthropy was inextricably bound up with the movement to imagine public schools as a private market. Conversations with Janelle Scott and Phillip Kovacs, both of whom have done early scholarship on venture philanthropy, were invaluable to orienting me in the right direction on this project. As well, Frederick Hess's terrible collection of neoliberal pro-venture philanthropy essays in *With the Best of Intentions* proved a very valuable resource in grasping the extent of what is wrong with venture philanthropy. I only hope my book will not be as useful to that cabal of privatization advocates as theirs has been to me.

A number of people were extremely generous with their time and thought. I am tremendously grateful to Henry Giroux (in whose series with Susan Giroux this book appears) who not only provided crucial insight on organization and ideas but also co-wrote the response to Arne Duncan's selection with me for truthout.org that appears as the coda here. Palgrave Macmillan editors Julia Cohen and Samantha Hasey provided valuable suggestions and were a pleasure to work with. Robin Truth Goodman once again talked through arguments and ideas and sacrificed many a brain cell, undertaking extensive editing of the manuscript. Again in appreciation for her dedication and hard work, I attribute any deficiency in the manuscript to her alone. I am grateful to Holly Christie who did extensive and valuable work analyzing tax forms of the venture philanthropists and preparing the tables of data in the appendix. As well, graduate assistants Moira O'Shea and Liz Fyffe did vast amounts of extremely valuable research for me. Thanks for the always thoughtful ongoing exchange to Enora Brown, Stephen Haymes, David Gabbard, Trevor Norris, Wayne Ross, Sandra Mathison, Pepi Leistyna, Chris Robbins, Pat Hinchey, Deron Boyles, and Kristen Buras, and Tom Nelson. Thanks also to Al Lingis, Noah Gelfand, Rob Isaacs, Kevin Bunka, and Chris Murray.

Great thanks to artist Diem Chau the photograph of whose carving appears on the cover.

# INTRODUCTION

Although educational philanthropy accounts for just a fraction of educational spending in the United States, its institutions have recently acquired disproportionate influence and control over educational policy and practice. Long-standing passive giving by such institutions as the Carnegie, Ford, and Rockefeller Foundations has receded as active players such as the Bill and Melinda Gates Foundation (the largest philanthropy in human history), The Broad Foundation, and the Walton Family Fund exert far more direct influence in multiple areas of education. *The Gift of Education* dissects the new "venture philanthropy," considering the educational and social implications of private donors, leading reform to K-12, teacher education, and higher education.

The new philanthropy is at the forefront of a right-wing movement to corporatize education at multiple levels. That is, venture philanthropy (VP) contributes to both the privatization of public schooling as well as the transformation of public schooling that is based on the model of corporate culture—from voucher schemes to charter schools to the remaking of teacher education, educational leadership, and classrooms. Educational philanthropy that appears almost exclusively in mass media and policy circles as selfless generosity poses significant threats to the democratic possibilities and realities of public education. Though focused on educational philanthropy, this book concerns much broader questions about the ideological agendas of philanthropic giving, the educational and political implications of the privatization of public governance, and the role of public money in financing and subsidizing private control of public services. Venture philanthropists are involved in all aspects of educational reform, policy, and practice from finance to administration, from pedagogy to curriculum.

Venture philanthropy differs markedly from the previous educational philanthropy that was dominant throughout the twentieth century, including large donors such as The Carnegie Corporation,

The Rockefeller Foundation, and The Ford Foundation. These traditional philanthropic endeavors defined giving through a sense of public obligation. In the traditional view, the industrialist gave back some of the surplus wealth that he had accumulated. Carnegie's *The Gospel of Wealth* codifies this perspective that its advocates described as "scientific philanthropy." As critics such as Robert Arnove, Joan Roelofs, and others have argued, the early educational philanthropy played a distinctly conservative cultural role in supporting public institutions in ways that were compatible with the ideological perspectives and material interests of the captains of industry rather than of the workers of coal, steel, oil, or automotive production. This labor created the surplus wealth that was first extracted and then given for universities, museums, libraries, and trusts. As Slavoj Zizek emphasizes, to *give*, the capitalist must first *take*.[1] Public subsidies through tax incentives not only encouraged but financed such public works to be developed and designed by fiscal and cultural elites rather than by the broader public.

Although educational philanthropy played a hegemonic role throughout the twentieth century, it was hardly unified in its approaches, and it offered funding for a wide variety of initiatives and projects that were not restricted to the conservative side of the political spectrum. There was a distance between the donors and the end uses made of the donated money in education—once given, the money was not closely controlled and directed in its uses. "Scientific philanthropy," though beholden to a logic of cultural imperialism,[2] was marked by a spirit of public obligation and deeply embedded in a liberal democratic ethos.

Venture philanthropy departs radically from the age of "scientific" industrial philanthropy. VP is modeled on venture capital and the investments in the technology boom of the early 1990s. VP not only pushes privatization and deregulation, the most significant policy dictates of neoliberalism,[3] but it is also consistent with the steady expansion of neoliberal language and rationales in public education, including the increasing centrality of business terms to describe educational reforms and policies: choice, competition, efficiency, accountability, monopoly, turnaround, and failure. Likewise, VP treats giving to public schooling as a "social investment" that, like venture capital, must begin with a business plan, involve quantitative measurement of efficacy, be replicable to be "brought to scale," and ideally "leverage" public spending in ways compatible with the strategic donor. Grants are referred to as "investments," donors are called "investors," impact is renamed "social return," evaluation becomes "performance measurement," grant-reviewing turns into "due diligence," the grant list is renamed an "investment portfolio," charter networks are referred to as "franchises"—to name but some of the recasting of giving on investment. Within the view of VP, donors are framed as both entrepreneurs and consumers while recipients are represented as investments. One of most significant aspects of this transformation in educational philanthropy involves the ways that the public and civic purposes of public schooling are redescribed by VP in distinctly private ways. Such a view carries significant implications for a society theoretically dedicated to public democratic ideals. This is no small matter in terms of how the public and civic roles of public schooling have become nearly overtaken by the economistic neoliberal perspective that views public schooling as principally a matter of producing workers and consumers for the economy and for global economic competition.

Although educational philanthropy and VP in particular represent a very small portion of the roughly $600 billion annual expenditure on education in the United States, VP has a strategic aim of "leveraging" private money to influence public schooling in ways compatible with the following: long-standing privatization agendas of the political right; conservative think tanks such as the Heritage Foundation, the Hoover Institution, and the Fordham Foundation; corporate foundations such as ExxonMobil; and corporate organizations such as the Business Roundtable and the Commercial Club of Chicago. The central agenda is to transform public education in the United States into a market through for-profit and nonprofit charter schools, vouchers, and "scholarship" tax credits for private schooling or "neovouchers." Venture philanthropies such as New Schools Venture Fund and the Charter School Growth Fund are being financed by the large givers and aim to create national networks of charter schools, charter management organizations, and educational management organizations (EMOs). These organizations are explicit about their intent to transform radically public education in the United States through various strategies. Working along these lines, the venture philanthropists also coordinate with large urban school districts and business groups to orchestrate such plans as New York's New Visions for Public Schools, Chicago's Renaissance 2010, and similar mixed-income schools and housing projects in Portland, OR, and Boston, MA, and elsewhere. These organizations coordinate the privatizations of schooling and housing and gentrify coveted sections of cities. They aggressively seek to reimagine teacher education through online and onsite initiatives and educational leadership on the model of the MBA. The key players of VP in education—including but not limited to such leaders as Gates, Walton, Fisher, and Broad—are able to exercise influence disproportionate to their size and spending power through strategic arrangements with charter and voucher- promoting organizations, think tanks, universities, school districts, and schools. The seed money that underfunded schools desperately seek allows the venture philanthropists to "leverage" influence over educational policy and planning, curriculum and instructional practices, and influence the very idea of what it means to be an educated person. Though the implications for educational reform are vast, there has been scant scholarship on VP in education.[4]

While this book provides the first comprehensive criticism of VP, it does not stop with criticism. It also provides a way of reframing multiple educational issues, policies, and practices by drawing on the literature of gift exchange that has largely remained outside the field of education. Drawing on literature from Marcel Mauss and other theorists of gift exchange and "general economy", *The Gift of Education* aims to turn the current neoliberal economism of U.S. education on its head by raising fundamental questions of what it means to give an education, who owes whom an education, and how the material and symbolic exchanges of knowledge and meaning-making involve expenditure, excess, consumption, and sacrifice. While neoliberal education aims to reduce public education to its possibilities for market economics, a general economy of education aims to expand the possibilities of education to all aspects of public and private life and also social relations. While the second, third, and fourth chapters criticize the multiple dimensions of the neoliberal venture philanthropy agenda, the last chapter suggests an alternative not only to the VP program but also to both neoliberal education and the theoretical assumptions of economism informing it and informing the unfortunate resurgence of economistic Marxism in education.

The first chapter foregrounds the illustration and criticism of VP in education by offering a broad overview of the corporatization of public schools including the economic, political, and cultural dimensions of it. This chapter explains multiple aspects of the corporatization of schools including the rise of neoliberal ideology,

transformations in educational and political governance, the fostering of consumerism, and the problems that school corporatization
poses to public and critical forms of schooling ideally suited to a
democratic society. The next four chapters detail how VP fosters
multiple forms of school corporatization including privatization,
charter schools, vouchers, grant competitions, the importation of
corporate culture in schooling, and the remaking of educational
governance, teacher education, and educational leadership.

The second chapter begins with examples of the various corporatization projects underway by venture philanthropists, including
the school projects of Gates, Broad, Walton, and Fisher as well as
the introduction of corporate culture reforms like paying students
for grades and paying teachers bonuses for test scores. It maps the
web of nonprofit and corporate foundations, local tiers of grassroots organizations, key players, and associations, showing how the
reach of VP extends to nearly all aspects of public schooling. This
chapter details the transformations in financial and political governance accompanying the rise of VP. It contends that the new philanthropy participates in a "circle of privatization" whereby public
financial subsidies to private individuals and businesses encourage
private giving that influences, leads, and directs public educational
initiatives. These financial redistributions, in turn, redistribute
public governance in privatized ways. The chapter explains how it is
that the public pays to give up control over its own institutions, and
how such a trend needs to be understood in terms of broader political battles.

This chapter details the way venture philanthropists are involved
in multiple educational reform initiatives that are framed through
the neoliberal idealization of competition, choice, efficiency,
accountability, and performance-based achievement. Such market
metaphors stand in for evidence of meaningful educational improvements while making privatization and deregulation seem naturally
beneficial by associating them with the unquestioned superiority of
business practices. What is being accomplished in the name of excellence and school improvement is a largely long-standing right-wing
agenda of smashing teachers unions and putting in place for profit
and nonprofit charter schools, vouchers, culturally conservative
pedagogical models dependent upon a cultural deficit logic, as well
as the coordinated taking of coveted real estate through the privatization of public housing and schools. It examines how the venture
metaphor relies heavily on intersecting metaphors of racial, gendered,

and market discipline. The chapter includes discussion of how VP claims to empower communities by expanding choice while actually undermining local community control. As well, it examines the effort of VP to "replicate" allegedly successful school models and to "scale up" charter schools into "franchises," revealing how the very proponents of replication declare it a myth used to achieve particular ends, namely expanding charter networks, removing unions, and furthering particular school designs. What is at stake in market-based school policies is the threat they pose to the public and civic possibilities of public education.

The third chapter explains what is new and different about VP from the age of scientific philanthropy of Carnegie and Rockefeller. It recounts the historical transformation of educational philanthropy in terms of the reworking of the social contract and the redefined notion of the public.

This chapter lays out the shift in the discourse of philanthropy away from public language and referents and toward the language and logic of investment. VP needs to be understood within the expansion of neoliberal ideology, generally, but also of neoliberal education in the early nineties. I show as well how the private language of venture capital behind the technology boom of the nineties was exported to the public and nonprofit sectors. The chapter discusses how the redefinition of educational philanthropy has been tied in multiple ways to upward redistributions in wealth and expanded inequalities in wealth and income in the "second guilded age." What is at stake in such an understanding is grasping how the new VP participates in a shift from state and public forms of social and self-regulation to individualized and market-based forms of social and self-regulations in an increasingly dual economy and society.

The fourth chapter focuses on the Eli Broad Foundation's initiatives to transform educational leadership as part of its urban education reforms including its educational leadership preparation initiatives, the Broad prize, and database projects. It illustrates how the VP reforms are united by not only a corporate vision for education but a view of knowledge, curriculum, and culture that undermines public, intellectual, and critical forms of schooling more conducive to democratic culture.

The fifth chapter explores the reach of the new philanthropy in corporatizing teacher education and higher education. The chapter maps the terrain of teacher education reforms considering liberal professionalization, neoliberal deregulation, and critical versions of teacher preparation. The chapter criticizes VP's push for the neoliberal deregulation agenda and its narrow definition of good teaching defined by test scores and the threats it poses to public forms of teacher education. The chapter examines the Gates Foundation's newfound emphasis on graduation rates and college preparation in relation to both the radically raised cost of higher education and the structural inability of the global economy to accommodate the number of graduates for which these projects aim. It also considers these initiatives in the context of the Obama administration's post-political practicalism with regard to education.

The sixth chapter discusses how the literature on gift exchange offers a new way to think about educational philanthropy that is radically different from that of the venture capital model. The problem with the venture capital metaphor that grounds VP is not merely one of how it shifts the political, civic, and citizenship dimensions of public schooling onto the private terrain of the market. Public school corporatization generally and VP suggest a particular kind of economy—that is, restricted economy. I begin by criticizing both neoliberal "right" economism and Marxist left economism. I then use Marcel Mauss's *Essay on the Gift* to propose alternatives to thinking about educational obligation, provision, policy, and practice. Mauss's theory opens the possibility for rethinking educational reform guided by "general economy" rather than "restricted economy." Such reframing tears educational discourse from its contemporary dominant reduction to business possibilities and inextricably links educational policy, practice, and provision to the whole social order through the registers of material and symbolic exchange. It also stands as a better alternative to the "new old" Marxist educational thought that has failed to learn from the extensive scholarship that criticizes the limits of economic reductionism and that inadequately theorizes politics and culture while offering a very narrow view of economy and human possibilities. This final chapter proposes a renewed and transformed "educational contract" and calls for rethinking educational obligation in order to build critical public democratic schooling as a means to achieving a critical democratic society.

These chapters are followed by a coda and an appendix. The coda, "Obama's Betrayal of Public Education? Arne Duncan and the Corporate Model of Schooling" was co-written with Henry A. Giroux for Truthout.org, and it appeared online on December 17,

2008 shortly after the announcement of Obama's selection of Chicago's Chief Education Officer for Secretary of Education. This article criticizes Duncan for his record in Chicago in terms of both the expanded repressive policies and aggressive implementation of school corporatization. The article was very widely distributed online and garnered requests for numerous media interviews from the authors. It stood out because it represented not only an early criticism of the rightward tack of the early Obama administration but because it broke with the largely conciliatory response to Duncan's selection by the teachers unions who suffered under Duncan's leadership in Chicago. The teachers unions and a number of liberal and progressive education authors expressed relief that Obama had not selected the even more anti-union and corporate reforms-oriented possibilities of DC superintendent Michelle Rhee or New York City schools Chancellor Joel Klein. For this book, the "Obama's Betrayal" article is a valuable inclusion because not only does it contribute a brief criticism of the new secretary's and new president's national approach to school reform derived from the business-designed Chicago plan but also because Duncan's corporate approach is lined up with the approach of the major venture philanthropists—especially Gates and Broad.

As this book goes to print, NBC has launched a television series/Microsoft advertising campaign called "The Philanthropist." The philanthropist is a British-accented, super-rich commodities trader who goes on James Bond-style missions to poor countries. Battling natural disasters, rebel armies, and the like in order to carry a box of vaccine to save a child, the philanthropist works through his own psychological trauma of having lost his son and consequently his marriage. While watching the show, one has difficulty imagining elderly financier George Soros or a sweatered Bill Gates doing what the main character does: trudging through the jungle, getting bitten by rattlesnakes, having an orgy with a group of female Nigerian prostitutes, or saving a child from flood waters. But the crucial idea in the show comes through loud and clear in the commercials. The show is heavily sponsored by Microsoft's web browser Bing, and the commercials for the web browser are almost seamlessly integrated into the show's content such that the search engine in the pilot episode appears as the key tool for the philanthropist's aids to help him do his good works. The show appears as an extended Microsoft commercial. The point not to be missed is that Gates's Microsoft is the truly generous force and the real hero rather than the cartoonish main character behind the saving of the Nigerian kid, the village, and so on. Moreover, the commercials implicitly make the point that the good works of the Gates Foundation are inextricably bound up with the profit-seeking and monopolistic behavior of the massive Microsoft. More importantly yet, the show and commercials instruct the viewer about how only the private sector, the super-rich philanthropist/capitalist and the corporation, rather than the public sector, including inevitably corrupt governments and nonexistent social movements, can be relied upon to solve public problems. Remarkably "The Philanthropist" registers the destructive role of the financial speculation of its main character in destabilizing the public sector in poor nations only to suggest that there is nothing to be done about this other than to rely on the exceptional philanthropic heroism of the main character supported by the real hero, the for profit technology of the Microsoft search engine.

The television show/Microsoft advertisement is a corporate-produced example of public pedagogy that teaches viewers how to understand for-profit technology, philanthropy, and public problems such as poverty, disaster, and war in ways that are compatible with a private sector view of the social and the individual. As such, it articulates with broader public discourses and corporate ideology. It also shares a broader social vision with VP in education that is at odds with critical views about the possibilities of public education. In the VP view, schooling serves a largely economic role of offering the individual inclusion into a corporate-dominated economy. Despite the rhetoric of "citizenship" found on the websites of Gates and Broad foundations, these venture philanthropies have a view of the role of schooling for citizenship and individual agency defined principally by economic inclusion and consumerism—a view of citizenship and agency at odds with questioning and collectively transforming broader oppressive structures and forces. The venture philanthropists' approach to schooling emphasizes standardized testing, standardized teaching, and the standardization of curriculum at the expense of an emphasis on individual and social contexts, the relationship between knowledge and authority, and the possibilities for acts of interpretation, dialogue and debate to form the basis for social intervention and social transformation. As the Microsoft-sponsored show highlights, agency is reserved for the exceptional individual (and the staff of lackeys who he abuses) who has successfully negotiated the private sector. For the venture philanthropists, schooling represents an opportunity to consume and regurgitate the official knowledge deemed by experts to be of universal value. Success in this domain can then be exchanged for further educational honors and advancement which can then be exchanged for opportunities to compete economically. In concert with the VP view of schooling and knowledge, the advertisement/ show illustrates the use of the information-gathering technology by the abused workers who scramble to collect knowledge in the service of the boss/philanthropist and his myopic projects. In the pilot episode, the philanthropist's project of addressing the public health effects of a hurricane and political conflict by bringing a single box of vaccine to a single village in Nigeria is downright idiotic, but his lackeys nonetheless use their Microsoft search engine to support this idiocy. The knowledge-gathering workers do Bing searches for the philanthropist to act alone. Knowledge is to be collected and then contributed to the top of the hierarchy. His wealth is the guarantee of the philanthropist's authority to intervene. The advertisement/show celebrates this plutocratic approach to social problems and learning. The show inadvertently highlights the anti-democratic tendencies of relying on the private super-rich individual, private for-profit technologies, and private financial speculation to address what are public problems that need to be address collectively. As the first chapter of this book elaborates, critical democratic approaches to schooling differ significantly from the corporate approach represented by venture philanthropy.

The VP approach to schooling like the advertisement/show largely imagines the possibilities of individual agency not through collective citizen participation but rather through consumerism. For the venture philanthropists, hierarchical, disciplinary, anti-intellectual, and positivistic school reform is justified on the basis on the promise of consumption. If the student works hard in school, then she can get a job and actualize herself as an individual by buying lots of stuff. The show "The Philanthropist" merges together the act of foreign intervention with the use of a technology that is, like most search engines, used largely for shopping. Like Bennetton's nearly two decades-old United Colors of Bennetton[5] advertising campaign and the current expanding rhetoric of "corporate responsibility", "The Philanthropist" suggests that consumer activity, using an advertising-based web browser, is the route to social change. This defines political engagement through the economic register of capitalism by framing civic engagement as consumption. It also wrongly suggests that the responsibility for social improvement ultimately lies with the least economically and culturally powerful social unit, the individual consumer rather than with the most powerful one, the corporation which has not only the most economic and meaning-making power but also a vested interest in doing whatever generates the most profit, socially destructive or not. In this sense, corporate "social responsibility" ought to be understood primarily as a corporate tactic for redistributing responsibility for externalities (the destructive side effects of profit seeking behavior) from themselves to everyone else.

The merged program content and advertising of "The Philanthropist" generally mirrors the merged web content and advertising of the internet, but it also adapts to the problem for advertising-based mass media posed by the expanding use of technologies like "tevo" and web-based television viewing, both of which allow users to evade commercials. "The Philanthropist," by merging advertising content with program content, really exemplifies the VP view that for-profit activity and good works ought to be the same thing (a point frequently and loudly made in interviews by Bill Gates, Bono, and Bill Clinton) and that philanthropy ought to be modeled on and emulate venture capital. In addition to denying the frequency of conflicts between profit accumulation and ethical intervention (like the profits of massive pharmaceutical companies threatened by free drug distribution for curable disease in the global south), this merging of for-profit and philanthropic activity necessarily eradicates both the distinction between public and private sector and the labor exploitation and financial speculation that forms the basis of wealth accumulation in a capitalist economy. While a corporate cultural product such as "The Philanthropist" plays a powerful pedagogical role of educating citizens about how to think about the self and community, economics and politics, ethics and progress, it is crucial to point out that the school reform championed by the venture philanthropists like Gates, Broad, and Walton offer no way to critically analyze such representations and the visions and values that they propose. Venture philanthropists push standardized testing and standardization of curriculum at the expense of critical pedagogical approaches that emphasize the crucial connections between the curriculum in schools and what is taught and learned outside of school including from other pedagogical sites such as mass media. Even though the Gates Foundation, which is the largest funder of VP school reform, is funded by the profits from new media (and public tax subsidy), it nonetheless in its educational projects utterly denies the pedagogical and public dimensions of its own media activities opting instead to promote retrograde educational practices reminiscent of Taylorism and industrial "scientific management." The kind of critical literacies necessary to comprehend and criticize the vast political and pedagogical force of new media are actively evacuated from what goes on in schools by the reforms of the venture philanthropists.

Educational practices inevitably contribute to particular visions for the future. Venture philanthropists imagine a corporate-dominated future in which the promise of consumption is the supreme promise. This is a vision in which individuals can only imagine a world made better by expanding markets and consumer goods. Not only is this vision of dubious sustainability economically and politically, but it is utterly unsustainable ecologically as unchecked economic growth as the supreme principle will assure ecological collapse and a resulting cascade of human disasters. As part of an effort at building a democratic culture and ethos, public schooling can provide the tools for criticizing existing realities and imagining new ones. However, such a critical approach to schooling stands starkly at odds with the VP vision of schools as mechanisms to churn out consumers and docile workers above all else. The chapters in this book aim to illustrate the limitations of VP's corporate approach to schooling and to suggest a renewed sense of the public in public schooling.

# FOREGROUNDING VENTURE PHILANTHROPY: THE CORPORATIZATION OF PUBLIC SCHOOLS

This book details a number of ways that VP stands to corporatize public schools from pedagogical approaches to curriculum, from transformations in governance and administration to changes to teacher and administrator preparation. This chapter foregrounds that discussion by providing a critical overview of the corporatization of schools, situating it in terms of broader political and educational trends. If, as this book contends, the corporatization of schools is a major problem with the business-oriented agenda of VP, then it is crucial to break down precisely what is meant by the corporatization of schools. This chapter serves as both a background for the following chapters that detail the multifaceted corporatization agenda of VP, and it serves as a starting point for the discussion of the problem of educational economism developed in the last chapter.

The corporatization of schools is part of a broader assault on public and critical education and the aspirations of a critical democracy. By the "corporatization of public schools," I mean both the privatization of public schools and the transformation of public schools on the model of the corporation. In what follows, I schematize corporatization in terms of economic, political, and cultural transformations. More specifically, I consider how the corporatization of public schools redistributes economic control and cultural control from the public to private interests. I argue that these intertwined redistributions of power undermine public democracy (the possibilities for the development of a more participatory and deeper democracy), just social transformation, and critical citizenship while exacerbating material and symbolic inequality.

Criticism of the corporatization of public education is predominantly restricted to the critical and radical political traditions.[1] For example, one tends to find criticism of corporatization framed by liberal writers as "business involvement in schooling" or by the more limited notions of "privatization" or "school commercialism". From right-wing perspectives, views on corporatization range from fiscal conservatives who champion privatization to cultural conservatives whose agendas are abetted by privatization and also to cultural and religious conservatives who worry about the ways in which business involvement in schooling threatens the traditions of schooling they support.[2] What distinguishes critical perspectives on corporatization is their focus on how privatization and the remaking of the school on the model of the corporation relates to broader social, political, economic, and cultural struggles.[3] From the critical perspective, the public school is a site and stake of struggle[4] for broader egalitarian social transformation. In each section, I will discuss how the criticism of corporatization from a critical perspective differs from liberal and right-wing views.

## THE EXPANSION OF CORPORATE POWER

While the institution of the modern corporation can be traced back to the origins of the European colonial expeditions of the late fifteenth century and their charters to exploit the new world (charters that initiated genocide of the indigenous populations of the new world as well as the genocidal African slave trade), the modern limited liability corporation dates back more recently to the mid-nineteenth century. Chartered stock corporations began historically to raise capital for public projects. The earliest corporations were highly unstable with spectacular financial implosions. Prior to the invention of limited liability, corporate collapse resulted in the financial responsibility of each shareholder for the project's debt. The advent of limited liability insulated the investing shareholder from devastating financial risk, paving the way for greater and greater levels of economic investment and development projects. However, limited liability also limited the responsibility for financial or socially devastating effects of investments from the individual investor. Once the stated mission of the corporation was accomplished, the corporation's charter would end.

New projects required new corporate charters from the public. By the late nineteenth century, states were battling to lure corporations and capital, and this race resulted in the expansion of incorporation rules, the loosening of controls over mergers and acquisitions, and the allowance of a company owning stock in another. The 14th amendment to the constitution was also reinterpreted such that the legal status of the stock corporation would be, in the eyes of the law, treated as a person. This expanded the rights of the corporate entity. By the early twentieth century, corporations gained further power In fact, professor and attorney Joel Bakan, author of *The Corporation* (the book was made into an excellent film by Mark Achbar), contends that if the corporation is to be legally considered to have the rights of a private individual, then we should consider what kind of person the corporation is.

How do we judge the character of an oil company that causes vast environmental devastation to cut costs, or an automaker who lobbies against airbags for decades causing untold deaths, or an agriculture company that genetically modifies crops in ways that threaten biodiversity and tries to patent life itself, or a fast food company that willingly destroys precious irreplaceable rainforest in South America for cattle grazing land to make burgers? How do we judge the character of any of these companies when they "partner" with schools because they know that they can make lifelong brand impressions on this captive audience?

Referring to the DSM-IV, Bakan argues that the corporation is akin to a psychopath exhibiting such traits as: socially damaging behavior, disregard for the well-being of others, singularly self-interested behavior, irresponsibility, manipulative tendencies, grandiosity, lacking empathy, asocial tendencies, and an inability to feel remorse.[5] Bakan stresses that such a critical view of the corporation should not be based in moralism but rather in a recognition of what sort of institution results from the kind of laws that allow or even encourage such behavior. For example, corporations are legally required to maximize the financial returns of the corporation for shareholders even if this results in "externalities," that is, destructive social effects. Bakan emphasizes that the corporation is an artificial and human-made entity and could be redefined through collective political action and will. Others have called for the revocation of corporate charters when corporations cease to serve the social good.

The publicly traded and privately held corporation today stands as arguably the most powerful social institution, eclipsing the centrality of power held historically by the church and the state. This is not to say that the corporation has completely replaced either the church or the state as an ascendant hegemonic institution. However, the corporation has come to dominate nearly every social domain: agriculture, mass media and information, biological sciences, healthcare, energy, and so on. As the ultimate corporate mission, *the capitalist imperative for the growth of financial profit at any cost*, is increasingly injected into all social domains, the social effects are felt everywhere. One effect is commodification: all social and individual things and values appear increasingly for sale. The commodification of the social world imperils collective public values and collective political agency as well as the public deliberation necessary for democratic governance. In nations theoretically dedicated to the promises of the liberal democratic political tradition, the imperatives for corporate profit have highly destructive effects: political campaigns are thoroughly based on advertising revenue and donations; political discourse is rendered all but meaningless as it is packaged into sound bites to fit between commercial messages; candidates are labeled as "electable" or "unelectable" by corporate media in ways that filter out candidates that pose a threat to corporate interests and values.

Corporations also have disproportionate hold over information and the representation of the social world in ways that undermine the possibilities for meaningful political deliberation to take on issues of public import or to enact radical change by transforming the rules of the game. Entire sectors of the economy such as mass media share interests with energy, heavy industry, and military corporations. NBC's parent company General Electric illustrates this clearly. Such interlocking corporate interests across sectors means that the mass media–led public discussion about, for example, how to understand and address global warming, must stay within the acceptable ideological framework—that is, the possibilities for action must not threaten corporate profit. (Interestingly, the power of a mass media company such as NBC to arrogantly admit this is made into comedy for profit on its television show *30 Rock*—this shift belies an ideological change as a decade earlier NBC censored a *Saturday Night Live* cartoon that highlighted the relationship. That we are all "in on" the open secret of corporate monopoly, stupidity, exploitation, and abuse positions us to experience it as inevitable and beyond challenge.) It is important to note that Microsoft and Bill Gates the leading and most powerful financial source for the largest venture philanthropy has consistently sought to monopolize the delivery of digital entertainment.[6] The corporate management and control of information in mass media shapes and limits public discourse and stands as a warning for what the increasing corporate control of public schools will do to the possibilities for schools to address matters of dire public import as well as schools' abilities to foster in students' investigative habits and critical dispositions. Such critical dispositions enable students to develop as critical citizens, linking subjects of study to broader historical struggles for power.

The hegemony of the corporation results as well in the global spread of the ideology of corporate culture. Corporatization, in addition to privatization, involves the corporate model of organization being applied to institutions that should not aim for the maximization of profit and growth. The corporate organization tends to be hierarchical if not authoritarian, sharing a form closer to the military than to that of participatory democracy. As public institutions including schools are remodeled on the corporation, their public and collective organization is replaced with authoritarian features. The ideology of corporate culture projects not only corporate models of governance but also fosters consumerism. Consumerism redefines individual and collective values such that possessive individualism, acquisitiveness, and market-based forms of association replace civic values, collective political aspirations, and ethical pursuits.

## CORPORATIZATION AND THE ECONOMIC CONTROL OF SCHOOLS

Although corporate involvement in public schooling goes back to the beginnings of public schooling,[7] the corporatization of public schools began in earnest in the early eighties as part of the rise of neoliberal ideology.[8] In the United States, public education has become increasingly privatized and subject to calls for further privatization while business and markets have come to influence or overtake nearly every aspect of the field of education. Privatization takes the form of for-profit management of schools, "performance contracting," for-profit charter schools, school vouchers, school commercialism, for-profit online education, online homeschooling, test publishing and textbook industries, electronic and computer-based software curriculum, for-profit remediation, educational contracting for food, transportation, and financial services, to name but a partial list. These for-profit initiatives include the steady rise of school commercialism such as advertisements in textbooks, in-class television news programs that show mostly commercials such as Channel One, soft drink vending contracts dominated by Coca Cola and Pepsi, sponsored educational materials that teach math with branded candy and sportswear, lessons in science and the environment by oil companies, and other attempts to hold youth as a captive audience for advertisers. The modeling of public schooling on business runs from classroom pedagogy that replicates corporate culture to the contracting out of management of districts to the corporatization of the curriculum to the "partnerships" that schools form with the business "community" that aim to market to kids.

## Public School Privatization

The EMO or Educational Management Organization focuses on managing schools for profit, 94% of which are charter schools. As of 2008–2009, at least 95 EMOs were operating in 31 states with 339,222 students and at least 733 schools with nearly 80% of students in school managed by the 16 largest EMOs. Major large companies include Edison Learning (62 schools), The Leona Group (67 schools), National Heritage Academies (57), White Hat Management (51), Imagine Schools, Inc. (76), Academica (54), the rapidly growing virtual online school company K12 (24), and Mosaica (33).[9] The largest EMO in terms of number of students, The Edison Schools (now Edison Learning), has been beset by numerous financial and accountability scandals that, as I explain in my book, *The Edison Schools: Corporate Schooling and the Assault on Public Education*, has less to do with corrupt individuals than with the impositions of privatization and the social costs of public deregulation.

Major privatization initiatives also include market-based voucher schemes allowed by the U.S. Supreme Court and implemented by the U.S. Congress in Washington, D.C. and in the gulf region following hurricane Katrina.[10] Education conglomerate companies such as Michael Milken's Knowledge Universe aim to amass a number of different education companies. These conglomerate companies hold a variety of for-profit educational enterprises, including test publishing, textbook publishing, tutoring services, curriculum consultancies, educational software development, publication, and sales, toy making, and other companies.[11]

In the United States, the ESEA law ("No Child Left Behind") has fostered privatization by investing billions of public dollars in the charter school movement, which is pushing privatization with over three-quarters of new for-profits being charter schools. "No Child Left Behind" (NCLB) also requires high-stakes testing, "accountability," and remediation measures that shift resources away from public school control and into control by test and text-book publishing corporations and for-profit remediation companies. For example, as The Edison Schools failed to profit financially as a "publicly traded" company, the company has shifted investment toward for-profit tutoring work through spin off companies Newton and Tungsten.

Despite a number of failed experiments with performance-contracting in the United States in the 1980s and 1990s, for-profit education companies and their advocates have continued to claim that they could operate public schools better and cheaper than the public sector. This claim appears counterintuitive: after all, how could an organization drain financial resources to profit investors and still maintain the same quality that the organization had with the resources that could be paying for more teachers, books, supplies, and upkeep?

Evidence appears on the side of intuition. To date, the evidence shows that it is not possible to run schools for profit while adequately providing resources for public education. This has been equally true whether the profit model is vouchers, charters, or performance-contracting. Nonetheless, the business sector, right-wing think tanks in and outside of academia, and corporate media continue to call for market-based approaches to public schooling. This has as much to do with ideology as with financial interest. For example, The Walton Family Foundation (the largest family-owned business in the United States is Wal-Mart) is the largest spender lobbying for public school privatization schemes in the form of vouchers and neo-vouchers (tax credits for private schooling). Assuredly, this has less to do with plans of the company to open Wal-Schools or interest in the public schools developing highly educated and thoughtful Wal-Mart "greeters" capable of union-organizing to break the anti-union commitments of the company than it does with the ideological beliefs of the Walton family that business works for them, so business should be the model for schooling.

Advocates of public school privatization rely on a number of arguments for their economic claims: (1) the larger the company becomes, the more it can benefit from "economies of scale" to save costs through, for example, volume purchasing and running schools across multiple states; (2) the private sector is inherently more efficient than the public sector because for-profit companies must compete with other companies and; (3) the private sector is more efficient because the public sector is burdened by regulations and constraints such as teachers' unions and the protections that they afford teachers that only get in the way of efficient delivery of educational services.

Proponents often justify commercialism and other for-profit initiatives on the grounds that they provide much needed income for underfunded public schools. However, even the business press by 2002 recognized that education is not good business: schools have too many variable costs for economies of scale to work; business would have to be spectacularly efficient to allow for quality and skimming of profits for executives while Enron, Worldcom, Martha Stewart, and The Edison Schools not to mention the collapsed banking and auto industries show just how inefficient business can be; far from regulations being a hindrance, they provide necessary protections against abuse of teachers' labor while providing financial transparency. As the largest ever experiment in privatization, The Edison Schools overworked teachers, misreported earnings, misreported test scores, counseled out low-scoring students, cheated on tests to show high performance to potential investors, and as they approached bankruptcy time and again, they revealed just how precarious and unaccountable that market imperatives can be when applied to education.

In the 1990s, the "cola wars" led to a race by soda companies to get vending machines and advertisements into schools. The subsequent public health crisis that includes unprecedented epidemic levels of obesity and type II diabetes in young children has given weight to multiple local struggles against school commercialism. School commercialism has grown steadily and taken a much larger form than simply soft drink vending.[12] Advertising in schools has reached new levels with sponsored educational materials, advertisements for Oreo cookies integrated into math lessons; advertisements lining school hallways, the sides of school buses, and scoreboards; electronic marketing; promotional contests (such as those run by Pizza Hut and Domino's); and Channel One, an advertising driven faux-news program launched by Christopher Whittle, the magazine entrepreneur who would go on to create the Edison Schools.

From a liberal and critical perspective, the privatization of public schools and the ideology of corporate culture need to be opposed.

For liberals, the goal is to strengthen public schools. Corporatization undermines the liberal promises of public schooling to make educated human beings and a thoughtful participating polity. From a liberal perspective, even though historically the public sector has failed to universally provide quality educational services equally to everyone, that still remains the goal. In this view, the expansion of the "best" schools, that is, those schools from class and racial privilege remain the model. Liberals such as Jonathan Kozol highlight the spending disparities between the rich predominantly white schools and the poor predominantly African American and Latino schools. Per pupil, rich schools get as much as four times more money than poor schools, while poor schools actually need more than rich schools. For liberals, the project of educational equality is very much defined by the equalization of educational resources toward the goal of inclusion—the equalization of educational opportunity is supposed to translate into economic and political opportunity for participation in existing institutions. For critical-ists, the defense of public schools is about defending the public sector toward the goal of critical transformation of the political and economic systems via the political and cultural struggle waged through civil society. In this sense, the cultural struggle to make public schools sites for the making of critical consciousness is crucial and is distinct from the liberal perspective. From a critical perspective schools can be democratic public spheres that can foster critical consciousness, democratic dispositions, and habits of engaged citizenry. From a critical perspective schools play a crucial role in producing subject positions, identifications, and social relations that can make radically democratic subjects committed to such projects as democratizing the economy, strengthening the public roles of the state, challenging oppressive institutions and practices, and participating in democratic culture.

## CORPORATIZATION AND THE CULTURAL CONTROL OF SCHOOLS

The cultural aspect of corporatizing education involves transforming education on the model of business, describing education through the language of business and the emphasis on the "ideology of corporate culture" that involves making meanings, values, and identifications compatible with a business vision for the future. The business model appears in schools in the push for standardization and routinization in the form of emphases on standardization of curriculum, standardized testing, methods-based instruction, teacher de-skilling, scripted lessons, and a number of approaches aiming for "efficient delivery" of instruction. The business model presumes that teaching, like factory production can be ever-more speeded up and made more efficient through technical modifications to instruction and incentives for teachers and students, like cash bonuses. Holistic, critical, and socially oriented approaches to learning that understand pedagogical questions in relation to power are eschewed as corporatization instrumentalizes knowledge, disconnecting knowledge from the broader political, ethical, and cultural struggles informing interpretations and claims to truth while denying differential material power to make meanings.

Business metaphors, logic, and language have come to dominate policy discourse. For example, advocates of privatizing public schools often claim that public schooling is a "monopoly," that public schools have "failed," that schools must "compete" to be more "efficient," and that schools must be checked for "accountability" while parents ought to be allowed a "choice" of schools from multiple educational providers, as if education were like any other consumable commodity. Shifting public school concerns onto market language frames out public concerns with equality, access, citizenship formation, democratic educational practices, and questions of whose knowledge and values constitute the curriculum.

As an offshoot of corporatization, market language and justifications for schooling eradicate the political and ethical aspects of education. For example, within the view of corporatization, students become principally consumers of education and clients of teachers rather than democratic citizens in the making who will need the knowledge and intellectual tools for meaningful participatory governance; teachers become deliverers of services rather than critical intellectuals; knowledge becomes discreet units of product that can be cashed in for jobs rather than thinking of knowledge in relation to broader social concerns and material and symbolic power struggles, the recognition of which would be necessary for the development of genuinely democratic forms of education.

School commercialism is the most publicized aspect of public school corporatization. This owes largely to liberal assumptions that commercialism taints the otherwise neutral and objective space of the school with business ideologies. From the progressive and radical traditions, such liberal horror at, for example, advertisements for junk food in textbooks is naïve because the school is already understood as a political "site and stake" in struggles for hegemony by different groups including classes, races, and genders.[13] Schools teach the knowledge and skills necessary for students to take their places as workers and managers in the economy. Skills and know-how are taught in ideological forms conducive to social relations conducive to the reproduction of relations of production. In the 1970s, this was dubbed the "hidden curriculum": students learn to be docile workers from teachers who emulate the boss; tests and grades prepare students for understanding compartmentalized often meaningless tasks and numerically quantifiable rewards that are extrinsic; earning grades prepares kids to work for money; school bells segment time in ways conducive to shift work while desks are arrayed with the teacher/boss at the big desk and the student/workers at the little desks...All of this suggests that the space of school is hardly free of capitalist ideology from the outset. As Henry Giroux has suggested,[14] the hidden curriculum is no longer hidden. As neoliberal ideology has resulted in the triumph of market fundamentalism in an overt fashion to all realms of social life, schooling has been remade on the model of the market.[15]

## Corporatization and Neoliberalism

Contemporary initiatives to corporatize public schools can only be understood in relation to neoliberal ideology that continues to dominate politics even as the basic tenets of it have been obviously discredited by the financial crisis of 2008 and the need for the public sector to bail out the private sector from unchecked deregulation.[16] Neoliberalism, a form of radical fiscal conservatism, alternately described as "neoclassical economics" and "market fundamentalism," originates with Frederic Von Hayek, Milton Friedman, and the "Chicago boys" at the University of Chicago in the 1950s. Within this view, individual and social ideals can best be achieved through the unfettered market. In its ideal forms (as opposed to how it is practically implemented), neoliberalism calls for privatization of public goods and services, decreased regulation on trade, loosening of capital and labor controls by the state, and the allowance of foreign direct investment. In the view of neoliberalism, public control over public resources should be shifted out of the hands of the necessarily bureaucratic state and into the hands of the necessarily efficient private sector.

Neoliberals seized upon the historical events of the collapse of the Soviet Union and the end of the cold war to claim that there could be no alternative to global capitalism. Within the logic of capitalist triumphalism, the only course of action would be to enforce the dictates of the market and expand the market to previously inaccessible places. As David Harvey writes, neoliberalism has been extremely successful at redistributing economic wealth and political power upward. For this reason, Harvey calls for understanding neoliberalism as a long-standing project of class warfare waged by the rich on the rest. Not only have welfare state protections and government authority to protect the public interest been undermined by neoliberalism, but these policies have resulted in wide-scale disaster in a number of places that have forced governments to rethink neoliberalism as it has been pushed by the so-called Washington consensus.

Originally viewed as an offbeat doctrine, neoliberalism was not taken seriously within policy and government circles until the late 1970s and early 1980s in Thatcher's United Kingdom and in Reagan's United States. Chile under Pinochet was a crucial testing ground for these ideals. The increasing acceptability of neoliberalism had to do with the steady lobbying for neoliberals by right-wing think tanks but also the right conditions including economic crises facing the Keynesian model and Fordism in the late 1970s. Neoliberalism has a distinct hostility to democracy. As Harvey writes,

> Neoliberal theorists are, however, profoundly suspicious of democracy. Governance by majority rule is seen as a potential threat to individual rights and constitutional liberties. Democracy is viewed as a luxury, only possible under conditions of relative affluence coupled with a strong middle-class presence to guarantee political stability. Neoliberals therefore tend to favour governance by experts and elites. A strong preference exists for government by executive order and by judicial decision rather than democratic and parliamentary decision-making. Neoliberals prefer to insulate key institutions, such as the central bank, from democratic pressures. Given that neoliberal theory centres on the rule of law and a strict interpretation of constitutionality, it follows that conflict and opposition must be mediated through the courts. Solutions and remedies to any problems have to be sought by individuals through the legal system.[17]

In education, neoliberalism has taken hold with tremendous force, remaking educational common sense and pushing forward the privatization and deregulation agendas. The steady rise of all of the reforms and the shift to business language and logic mentioned in the earlier sections can be understood through the extent to which neoliberal ideals have succeeded in taking over educational debates. Neoliberalism appears in the now commonsense framework of education exclusively through presumed ideals of upward individual economic mobility (the promise of cashing in knowledge for jobs) and the social ideals of global economic competition. The "TINA" (There Is No Alternative to the Market) thesis that has come to dominate politics throughout much of the world has infected educational thought as the only questions on reform agendas appear to be how to best enforce knowledge and curriculum conducive to national economic interest and the expansion of a corporately managed model of globalization as perceived from the perspective of business. What is dangerously framed out within this view is the role of democratic participation in societies ideally committed to democracy and the role of public schools in preparing public democratic citizens with the tools for meaningful and participatory self-governance. By reducing the politics of education to its economic roles, neoliberal educational reform has deeply authoritarian tendencies that are incompatible with democracy. As the only concern becomes one of the efficient enforcement of the "right" knowledge, critical engagement, investigation, and intellectual curiosity appear as impediments to learning, and teachers are deskilled deliverers of prepackaged curricula prohibiting their potential as critical intellectuals. Educational language has been overrun with neoliberal terms that undergird the framing of educational issues through the ideal of "achievement," "excellence," and "performance-based assessment." These nebulous terms falsely presume agreement over what is meant by these goals.

Neoliberals have sought to lay claim to the meaning of "democracy" as well. John Chubb and Terry Moe wrote in their 1991 book *Politics, Markets, and America's Schools*[18] that the market better facilitates democracy than the political involvement of the public. For Chubb and Moe, the bureaucratic strictures of the public sector disable democratic administration while the market facilitates it. As liberal scholar Jeffrey Henig pointed out in his book *Rethinking School Choice*,[19] Chubb and Moe's conception of democracy understands democracy through administrative procedure rather than collective public enactment of shared political and ethical commitments and visions.

All critical views on corporatization of public schooling share certain basic assumptions:

1. We live in a fundamentally unjust social order: politically, economically, and culturally.
2. Public schools either function hegemonically to contribute to the reproduction of this unjust social order or public schools can be sites for counter-hegemonic struggle.
3. What goes on in schools (school model, curriculum, pedagogy) matters as to whether public schools function predominantly as hegemonic or counter-hegemonic sites.
4. Critical educational practices can produce critical subjects.
5. Collective social transformation requires critical subjects who can theorize and act—engage in praxis.

How does corporatization threaten the critical vision? Liberal and conservative perspectives on public schooling operate through accomodationism. That is, they presume that we live a fundamentally just social order and that the role of public schools is to accommodate students to that order. While historically and presently state-adminstered public schooling functions hegemonically, both privatization and the ideology of corporate culture deepen the conservatizing tendencies of public schooling rather than unsettling them. In part, this is due to the ways that corporatization reworks public education through the discourse of *economism*.

Corporatization conflates the public and private purposes of public schooling, treating schooling like a for-profit business. While state-administered public education is grounded in the model of the industrial economy, public schooling prior to the advent of corporatization retained secular humanist ideals as well as reference to broader public purposes of public schooling. Corporatization as an expression of neoliberalism reduces the purposes of schooling to economic ends. On an individual level, the school becomes a means to upward economic mobility for the individual. On a national level, the school becomes a player in global economic competition. The economism prosthletized by corporatization produces a version of the social world in which the capitalist economy and the individual's submission to its dictates becomes the organizing principle behind learning and teaching.

Part of what is at stake in the corporatization of schools is the diminishment of the public sphere. Some people like Stephen Ball have recently suggested that the distinction between public and private in education is too blurry and complex to allow a meaningful distinction between public and private or to justify defending public schools and that, it is not clear what public values in education might be.[20] Ball appears alternately critical of and sympathetic to projects that allow corporations to contract with the state to run schools. There are at least four clear ways that those committed to democratic education must understand how public control differs from private control. I use The Edison Schools[21] as an example in each case to illustrate this. The Edison Schools company is an apt example to use as charter school promotion, has been a central campaign of venture philanthropists and has been emphatically embraced by Obama. The Obama administration's emphasis on the public funding of the expansion of charter schools will increase the prevalence of for-profit management companies like Edison contracting with districts to manage schools for profit. One-half of The Edison Schools are charter schools and the majority of new for-profits are charter schools. Consider these ways that public control differs from private control:

1. *Public versus private ownership and control*: Edison is able to skim public tax money that would otherwise be reinvested in educational services and shunt it to investor profits. These profits take concrete form as the limousines, jet airplanes, and mansions that public tax money provides to entrepreneur and majority owner Chris Whittle. These profits also take symbolic form as they are used to hire public relations firms to influence parents, communities, and investors to have faith in the company. This is a parasitical financial relationship that results in the management of the schools in ways that will maximize the potential profit for investors while cutting costs. This has tended to result in antiunionism, the reduction of education to the most measurable and replicable forms, assaults on teacher autonomy, and so on. There is no evidence that the draining of public wealth and its siphoning to capitalists has improved public education or that it is required for the improvement of public education. If the state is going to use privatization as a tool (as the advocates of the Third Way in the United Kingdom do), then they could exercise authoritative state action directly in ways that do not upwardly redistribute wealth or funnel such wealth into misrepresenting the public influence and effects of privatization.

2. *Public versus private governance*: There are numerous aspects of the transformation in governance accompanying privatization including the shift away from community governance, union governance, and the shift to business group governance. In Chicago, public schools are being closed under "Renaissance 2010" and reopened as for-profit and

nonprofit charter schools. Such schools are robbed of their community school councils, and business-dominated councils are installed. With Edison, decisions regarding the use of resources shift from community to a management team with a financial stake in particular outcomes. What is more, public tax money is brought in as profit and is then reinvested in public relations firms that lobby and influence the community (the public) to support Edison.

3. *Public versus private cultural politics*: Privatization affects the politics of the curriculum. A company like Edison cannot have a critical curriculum that makes central, for example, the ways corporatization threatens democratic values and ideals. While most public schools do not have wide-ranging critical curricula, the crucial issue is that some do, and most could. This is a matter of public struggle. Privatization forecloses such struggle by shifting control to private hands and framing out possibilities that are contrary to institutional and structural interest. The possibility of developing and expanding critical pedagogical practices are a casualty of privatization.

4. *Public versus private forms of publicity and privacy including secrecy and transparency*: Private companies are able to keep much of what they do secret. Edison could selectively reveal financial and performance data that would further its capacity to lure investors. Such manipulation is endemic to privatization schemes.

Collapsing public and private naturalizes public education as a private business despite fundamentally different missions.

As public schools are privatized, they are subject to a market-based logic of achievement in which knowledge is thought of as units of commodities to consume and regurgitate; it can be cashed in for grades; the grades can be cashed in for promotion; and the promotion can be cashed in for jobs and cash in the economy. The overemphasis on standards and standardization, testing, and "accountability" replicate a corporate logic in which measurable task performance and submission to authority become central. Intellectual curiosity, investigation, teacher autonomy, and critical thought, not to mention critical theory have no place in this view.

The charter school movement now being aggressively promoted by the Obama administration is seldom recognized as an aspect of corporatization. Yet, it typifies the social costs of the neoliberal ideals of deregulation and managerialism as they play out in education. Charter schools aim to minimize the "bureaucratic red tape" alleged to be responsible for the problems faced by traditional public schools. A business metaphor of efficiency is merged with a celebration of entrepreneurial experimentation to suggest that public regulations keep schools from being efficient and that the entrepreneurial spirit of the private sector is all that is needed. The neoliberal mantra of deregulation is applied to create public schools not subject to unions, with reduced administrative controls, and in many cases public over-sight. Although to date there is no evidence of charter schools being better than traditional public schools (and growing evidence to the contrary), charter schools weaken the public mission of universally good public schools and set the stage for further privatization. Once the control of public school gets devolved to local managers, it is merely a matter of swapping in a different manager. The central idea of "efficiency" defined by ever increasing test scores is the only way to understand quality in this view.

The Edison Schools again provide a ready example of the ways that privatization as a form of corporatization puts into place econo-mism and anti-criticality. As Edison was struggling to expand to become profitable, it had to show investors steadily improving test scores. The profitability of the company was contingent upon its con-tinued expansion to be able to achieve "economies of scale" in order to deliver more than the public schools while skimming out profit at the same time. The possibility of expanding the business and becom-ing profitable depended upon getting more investor capital. Investors needed to see evidence that Edison was superior quality to the public schools. Consequently, Edison put tremendous pressure on schools to achieve higher and higher test scores to show investors and to use in public relations. This resulted in reports of teachers cheating on tests and encouraging students to cheat on tests.

What is more, the tests became synonymous with educational quality. The possibilities for critical forms of education that engage with power relations, politics, and ethics are foreclosed when in conflict with the institutional interests of the company running the school. Put differently, will an Edison school ever include meaning-ful criticism of corporate power as part of its curriculum? Can it? Moreover, Edison and other educational privatizers target poor and working-class communities. That is, they target those communities that have been historically shortchanged by inadequate funding. These are students slated largely for the low paying end of the econ-omy. Critical curriculum and school models could provide the means for theorizing and acting to challenge the very labor exploi-tation that schools such as these prepare students to submit to. Edison does not target for privatization schools in economically and racially privileged communities. Privileged schools not only

benefit from success at capturing the bounty of public wealth, but
they also prepare students for the critical thinking necessary to take
management and leadership roles in the economy. Of course, critical thinking in the form of problem-solving skills is very different
from the kind of critical theorizing that would allow students to
comprehend the social and individual costs of their privilege and
learn to labor for something other than the corporate dream of
unfettered consumption.

In the progressive tradition, public deliberation on matters of
public importance is struggled over by citizens and groups. Culture
in the progressive tradition is to be interrogated rather than wor-
shipped or feared. What is common throughout the progressive
tradition is the idea that acts of interpretation become central to
acts of political intervention and participation. That is, in the progressive tradition, the meaning of democracy and the contents of
democracy as well as the contents of the culture are subject to interpretive struggle. The progressive tradition understands democracy
as dynamic rather than static, as shot through with multiple power
struggles, and as a quest and process rather than an achieved state
that must be fixed and held and protected from corruption. In the
progressive educational tradition, a democratic society requires citizens capable of not just functional literacy but also critical
literacies.[22]

Public schools are unique in that they hold the public potential
to foster such democratic dialogue and debate rather than being
reception centers for the knowledge, values, and virtues handed
down by self-proclaimed experts. Public school corporatization
threatens the possibility for public schools to develop as places
where knowledge, pedagogical authority, and experiences are taken
up in relation to broader political, ethical, cultural, and material
struggles informing competing claims to truth. Struggles against
these ideologies and their concrete political manifestations must
link matters of schooling to other domestic and foreign policies. It
is incumbent upon progressive educators and cultural workers to
imagine new forms of public educational projects and organize to
take back privatized educational resources for public control.

Although historically public education in the United States has
functioned to reproduce racial, class, and gender oppression, among
others, it has also been central to if not at the forefront of social
movements such as civil rights and grassroots multiculturalism.
Public schooling has also been open to ongoing experimentation,

tinkering, and response to intellectual movements across the political spectrum including good ones like progressivism and bad ones like scientific management. More importantly, beyond responding to social and cultural trends outside of schools, public schools themselves are sites of cultural production. The cultural politics of education do not go away. In other words, teachers as cultural producers are inevitably engaged in making meanings, values, ideologies, and identifications. The crucial questions are under what conditions and with what constraints do they do so. The sanction of commercialism, for example, produces commercial meanings and values, makes subjects as principally consumers, and undermines citizenship and the very notion of the public. However, school commercialism can be taken up critically by teachers to highlight the kinds of values, ideologies and interests represented by a particular product. Such analysis ought to include a focus on the material and symbolic interests embedded in the cultural text as well as analysis of what kinds of identifications and identities such commercial culture asks students to become. The possibilities of critical pedagogical enagagement with corporatization highlight the limitations of the liberal approaches to it. For example, liberal approaches to school commercialism end with the demand to keep public schools free of commercial content. Critical pedagogy offers the capacity to use commercialism to criticize the broader structures of power informing its very presence in the school.

Privatization schemes, however, materially limit what kinds of critical cultural production can be done. For example, when Disney runs schools in their corporate town, Celebration in Florida, a curriculum addressing what role Disney or ABC or the media monopoly plays in making public meanings is out of the question as is the likelihood of any serious questioning of the role of the corporation. The implications for a self-critical society are dire. Though public schools do often serve as ideological state apparatuses, they are nonetheless open to the possibility of being remade in democratic ways because ownership and control of such schools remain public and stay within the realm of public debate and oversight. This being the case then, the question is how to strengthen and further democratize a public system that needs to be understood as a crucial place for the making of critical democratic citizens, and as a base for democratizing the economy. What are the many ways that democratic cultural politics can be fostered within public schools to reinvigorate democratic culture everywhere? How can corporatization initiatives be reversed so as to reimagine public schools as the site of radical social transformation rather than the reproduction and entrenchment of existing relations of power? There is already an enormous defensive backlash against such anti-critical movements as the standardization of curriculum and the high- stakes testing regime. But progressives need to take the offensive by putting forward critical curriculum and approaches and pursuing concrete goals to take back public spaces. What should not be forgotten is that while the battle for critical public schools and against corporatization is valuable as a struggle in itself, it should also be viewed as an interim goal to what ought to be the broader goals of the left: redistribute state and corporate power from elites to the public while expanding critical consciousness and a radically democratic ethos.

The next four chapters details the ways that VP in education fosters the neoliberal corporatization agenda, from the shift in ownership and control of schools, to the transformation of school culture to influencing public policy, to the transformations of education leadership and teacher preparation, to enacting a vision of consumerism and undermining a democratic ethos and culture inside and outside of schools. It is imperative for progressives to challenge VP not only on the basis of efficacy understood through test scores or even costs. Rather, VP, as the next chapters make clear, represents an assault on the public possibilities of public education and the socially transformative potential schools have in a democratic society.

## 2

# THE TROJAN SCHOOL: HOW VENTURE PHILANTHROPY IS CORPORATIZING K-12

## INTRODUCTION

In the summer of 2006, Warren Buffet, widely regarded as the most successful investor in history, made international news when he announced that he would give $31 billion to the Bill and Melinda Gates Foundation. Buffett's gift not only comprised the bulk of his fortune, but it more than doubled the combined amounts that Andrew Carnegie and John D. Rockefeller donated to their foundations. The Gates Foundation stands as the largest philanthropy in history by far, valued in 2008 at $31.5 billion and with an annual budget of more than $3.5 billion. The Gates Foundation has been largely lauded in mass media as it funds health projects and champions reforms of public education. What has gone largely unnoticed by mass media and most scholars is that the Gates Foundation is the largest player in a fundamental transformation of educational philanthropy: it is setting the agenda for modeling public education in the United States on venture capital.

Despite its massive size and influence, the Gates Foundation is but one piece of a vast network of venture philanthropies in education. Though most often represented as generous givers to public schools or much needed leaders of reform, venture philanthropists in education do not necessarily aim to improve public education. Though often dressed up in the language of the public good, reform, improvement, or even citizenship, the literature[1] of the venture philanthropists documents a desire to destroy public education and replace it with privatized educational provision. The literature debates the strategies most effective to get the job done.

Venture philanthropists aim to transform public education in the United States by replacing public schools with privatized schools financed by vouchers and tax credits, replacing public schools with charter schools, replacing public districts with charter "franchises," and expanding corporate culture into all aspects of schooling. By "corporate culture," I am referring both to the school being reorganized on the model of the corporation as well as what Henry Giroux has referred to as "the ideology of corporate culture." Examples include not just the commodification of knowledge, but also the reshaping of identities in privatized and individualized ways such as the student as consumer of education. At the center of the agenda is the framing of this crucial public service through the market-based idealization of "competition" and "choice."

Venture philanthropy centrally aims to create an educational market across the United States. In this view, parents and students would be forced to shop for educational services, and public schools would be forced to compete against privatized schools for scarce tax dollars, scarce grants, and students. Venture philanthropists discuss using private foundation money to "leverage" the transformation of public schools and districts. They broadly aim to use foundation money to induce public administrators and teachers to transform the workings of public schools to facilitate their vision of privatization. What is largely represented in both mass media and educational policy literature as generosity, care, and goodwill is nothing short of a coordinated effort to destroy public education. That is, what appears as generosity and goodwill is in reality its diametrical opposite: the destruction of universal provision for public education, the foundation for deepening educational inequality rather than an attempt to remedy it, the production of a system primarily designed to benefit investors at the expense of the poorest citizens, and a worsening of the racial and gendered inequality that currently structures public schooling.

The United States spends roughly $500 billion a year on public schooling. However, as proponents point out, VP represents a fraction of educational spending—with foundation spending at roughly $1.5–2 billion as of 2005[2]. How could such a small portion of educational spending matter? As proponents such as Frederick Hess, Matthew Greene, and others argue, VP has the potential to transform both the financial administration and governance of public schooling nationwide. This transformation stands to redistribute control from teachers, parents, students, and communities to private foundations, for-profit and nonprofit organizations, business groups, and investors. Venture philanthropists debate how to best "leverage" private foundation dollars to influence and affect public school spending, administration, and school practices.

Turning public schools into privatized schools includes replacing public schools with charter schools to de-unionize them and allow for the subcontracting of for-profit educational management organizations, investing in administrative training that is based in business rather than public service, investing in teacher bonus pay tied to test scores, creating school reliance on grants from foundations and undermining political will for universal financing, and investing in school "replication" for schools that cohere with the privatizing vision of the venture philanthropists—school models that tend to be authoritarian, emphasizing heavy discipline, uniforms, and corporate culture.

This chapter begins by illustrating the hostility of venture philanthropists to public education and the centrality of the privatization agenda. I argue that VP produces a "circuit of privatization" whereby the most crucial aspects of the phenomenon redistribute educational governance and financial control upward in a highly anti-democratic direction. The sections that follow illustrate this by focusing on vouchers and tax credits, charter schools, and the expansion of corporate culture. This chapter illustrates how VP induces citizens to subsidize financially their own abdication of financial and administrative control over public schools. The discussion provides an overview of the rising influence of VP. I conclude by providing a rough map of the network of VP in education.

## KILLING PUBLIC SCHOOLS WITH KINDNESS

In 2004, Bill Gates appeared before the National Governors Association and gave a speech, a version of which was reprinted in multiple newspaper op-ed columns. Gates stated that, "Our high schools are obsolete. By obsolete, I don't just mean that they're broken, flawed or under-funded, although I could not argue with any of those descriptions. What I mean is that...even when they work exactly as designed, our high schools cannot teach our kids what they need to know..."

"This is an economic disaster," he said, and one that is ruining children's lives and "is offensive to our values."[3]

The other leading venture philanthropists have sounded the same alarm as Gates's declaration of the failure of public schooling and the threat that public education poses to the nation's economy. Even before Gates's speech, billionaire venture philanthropist Eli Broad, announced that "public education is in many ways in a crisis that we can no longer ignore...We risk not only a lower standard of living and a weaker economy...We're in danger of becoming a second-class nation. I see the stakes as incredibly high. We're headed in the wrong direction."[4]

For Gates, Broad, and the Walton family (of the Walmart fortune), who are the leading venture philanthropists in education, the "right direction" is to treat public schooling like a market, to make public schooling into a market. They share a set of assumptions about public education that can be generally summarized as follows[5]:

1. Public schools have "failed," and the public sector cannot be relied upon for "good" education.
2. Government is inefficient, and markets are efficient.
3. Government involvement threatens rather than facilitates personal liberty.

The declaration of the "failure" of public schooling forms the backbone of the venture philanthropists' school agenda. In my prior work, I have elaborated on the destructive implications of describing public schools and public goods generally as private goods.[6] I have emphasized that terms such as "failure," "choice," and "competition" (as well as "consumers," "efficiency," and "monopoly") are part of a broader long-standing neoliberal agenda[7] that extends far beyond education: misrepresenting public goods as private consumables, replacing the collective purpose of general welfare with the misguided terminology of profit accumulation, and portraying citizenship as consumerism.

Venture philanthropy in education needs to be understood as centrally an expression of neoliberal economic doctrine and ideology. At its most basic, neoliberal economic doctrine calls for privatization of public goods and services and the deregulation of state controls over capital, as well as trade liberalization and the allowance of foreign direct investment. As an ideology, neoliberalism aims to eradicate the distinction between the public and private spheres, treating all public goods and services as private ones. It individualizes responsibility for the well-being of the individual and the society, treating persons as economic entities—consumers or entrepreneurs, and it has little place for the role of individuals as public citizens or the collective public responsibilities of democracy. Within the purview of neoliberal ideology, the state can only be bureaucratically encumbered and inefficient, and the market naturally tends toward efficiency and effectiveness. Despite the antipathy to the state, neoliberals aim to shift the use of the state from its care-giving roles to its repressive ones.[8]

In the last two decades, neoliberal ideology has taken hold with a vengeance in education. This has involved describing public schooling as a business: students as "consumers," schools ideally needing to "compete" against one another, this competition driving up "efficient delivery," administrators described as "entrepreneurs" and schools needing to be "allowed to fail" "just like in business." "High stakes" standardized testing and standardization of curriculum have been utterly central to the neoliberal education agenda in part because of the ways it treats knowledge as a commodity to be produced by experts, delivered by teachers, and consumed by students. The critical and dialogic dimensions to learning and teaching are denied in this view that treats education as indoctrination of the "right knowledge." In the neoliberal perspective, this anti-critical view of knowledge and learning is labeled "student achievement." The metaphorizing of public education in the language of the market has confused the private enterprise of profit accumulation with the public and civic purposes of public education.

Within the neoliberal view of education, declarations of "failure" have more than one function. As a rhetorical strategy, they make it seem as though the fault for low scores has to do with the low merits of students and the underperformance of teachers rather than excessive standardization, financial pressures on school systems, over-strained parents, economic disadvantages, or the over-bureaucratization of knowledge. They thus redefine public schooling as private enterprise, and they naturalize private enterprise as the cure to public school "failings." As they conflate public and private sectors, they conceal how different levels of public investments result in different levels of educational quality and reflect historical inequalities in public investment.

The neoliberal declaration of a "failed system," which relies on the metaphor of business failure, is selectively deployed and is, racially and class coded. It is not leveled explicitly against rich, pre-dominantly white communities and public schools for which high levels of historical investment and the benefits of cultural capital have resulted in high achievement, traditionally defined. Rather, the declaration of "system failure" is leveled against the working class and the poor, predominantly nonwhite communities and schools. As such, it misattributes educational inequalities and short-comings to the public sector rather than to the private sector, which, in the United States, bears responsibility for them as funding is linked to wealth through property taxes and local funding while the private sector has historically played a central role in engineer-ing educational inequality.[9] For example, as Paul Vallas and then Arne Duncan in Chicago have pushed the neoliberal approach through first "reconstitution" and then the Renaissance 2010 plan, the endlessly repeated suggestion has been that the public sector has "failed," and now it is "time to give the market a chance." The market, however, got a chance when the business sector influenced and shaped school reform and policy in Chicago for over a hundred years prior to Renaissance 2010. The neoliberal solutions of union-busting, school privatization, the idealization of deregulation in the form of charter schools, the idealization of competition and choice, the business-led reform with events such as "Free to Choose, Free to Succeed: The New Market in Public Schooling," and the imple-mentation of "turnarounds" as seemingly innovative solutions actively deny the ways in which the public system has a long history of business-led engineered failure not to mention a history of unequal resource distribution by being tied to property wealth.[10]

Peter Frumkin, a prolific author on VP, highlights the link between the declaration of public school failure and blame for global economic competitiveness, by writing, "(O)ne of the most popular fields for venture philanthropy efforts has been K-12 edu-cation. Many business people see the failure of large parts of the public school system as a crisis that has the potential to erode America's long-term economic growth potential."[11]

So, when Gates and Broad declare the "failure" of public school-ing as potentially causing a broader economic crisis for the United States in the world, they are calling for a turn to the private sector to redress the problems that too much private sector involvement in education created in the first place. Moreover, these business tycoons misrepresent their desire for an educated workforce where workers would compete in the global economy as a universally valuable vision rather than a class-specific one that benefits the most those who own and control capital. Subjugating the public purposes of public schooling to primarily that of making competitive workers for the global economy presumes that the public interest is principally served by engaging in the global race to the bottom fostered by the neoliberal vision of trade deregulation and public sector privatization. Venture philanthropists openly talk about U.S. students ideally becoming workers who will compete for scarce jobs against workers from poorer nations. Values of worker discipline, docility, and submission to authority are injected into the corporate school vision as they represent the ideal of the disciplined, docile, and submissive workforce. This view of the national education system, serving the interests of capital in a global economy, is at odds with the public interest that would be better served by a critical pedagogy in which students develop the tools of social criticism and develop as critical intellectual citizens.

Right-wing foundations support and work in conjunction with venture philanthropists and share the vision of transforming public education into a private market. Chester Finn of the Thomas B. Fordham Foundation put it bluntly, "We support real programs such as charter schools and a privately-funded voucher-type program and various frustrating efforts to improve the public school system itself...What public education needs is to be forced to change...that force can come from the marketplace; from the *customer*, via *competition* from private schools, and charter schools and virtual schools and privately managed schools and home schools and much more."[12] The VP emphasis on "forcing" public schools to change misunderstands the motivations of teachers to teach, students to learn, and administrators to manage schools well. In this view, students and teachers are problems—obstacles standing in the way of enforcing the right knowledge by any means necessary. Teacher desires to foster student curiosity, and intellectual developments are irrelevant for the enforcement view which is championed by VPs and right-wing foundations alike. The Fordham Foundation funds VP initiatives and is funded by VPs. Likewise, Pete DuPont of the Lynde and Harry F. Bradley Foundation (a major funder of voucher projects) described the public school system as "awful," the worst thing the government does in America, and as "collectivism" that could be remedied by creating a market in education, and again by treating students and parents as consumers of private education. Similarly, Bruno Manno, who has served on the board of Fordham and a senior associate at the Casey Foundation, wrote in the Hoover Institution-published *Primer on America's Schools* that, "the present school enterprise is not just doing poorly, but is incapable of doing much better because it's intellectually misguided, ideologically wrong-headed, and organizationally dysfunctional."[13]

Finn and DuPont's statements belie not only neoliberal ideology with its values on privatization and deregulation but the longstanding project of rolling back social spending and the vestiges of the liberal welfare state to the days before Franklin Roosevelt's New Deal. In this market-fundamentalist view, "collectivism" signals an interpretation of public spending for social provisions such as education, healthcare, or housing as akin to communism and anathema to a market economy. In reality, the history of "scientific philanthropy," as I argue in the Introduction and expand in the next chapter, reveals support of industrial capitalists and their foundations for various welfare state protections in part as a defense against radical organizing and fears about the replacement of a capitalist economy with a socialist or communist one.

Rick Cohen's report for the Center for Responsive Philanthropy offers a thorough picture of the network of venture philanthropists and the privatization agenda they seek through tax credits and vouchers. Cohen is worth quoting at length on what he sees as the two ways of interpreting the intentions of the Walton Foundation, which is, by far, the largest VP working to privatize public education by expanding vouchers that use public taxpayer money to pay for private school tuition while defunding universal public schooling.

> It may be that the Walton support for public charter schools and public education in general is no more than rhetorical camouflage for a strategy to break public school teachers' unions and outsource the management and operations of K-12 education to private school operators as part of a Milton Friedmanesque kind of assault on the ability of the public sector to deliver quality education. However, a more reasonable approach is simply to take the Walton formulation at face value and recognize it as part of a well-capitalized and effective strategy to promote privatized alternatives to what conservatives see as a public school monopoly, mobilizing philanthropic and political capital to achieve a compelling and well thought out vision of K-12 education.[14]

Unfortunately, Cohen is right about both of these interpretations as they work synergistically to achieve the same ends. It is clear that the venture philanthropists do aim to break unions, outsource and commercialize management and operations, and they wage an assault on the ability of the public sector to deliver quality education. Their explicit aim is to force public schools to compete with private schools for scarce and finite public resources. Moreover, they aim to redistribute both the financial and operational control of schools to private interests. The New Schools Venture Fund (a major VP dedicated to expanding charter schools), for example, has a long-term strategy that involves first going after small school districts to capture a high percentage of the system as charter schools to then later be able to leverage its influence over the entire district. The intention of venture philanthropists to destroy the public system is compatible with both their faith in markets as necessarily and inherently superior to the public sector and their vision of replacing the public system with a private market.

In a sense, Cohen's dichotomy misses the fact that the intentions of the venture philanthropists are less important than the effects of their actions on public education—how it transforms not only the control and ownership of it, but the very meaning it might have and its role in a society committed to public democratic values. Writing of foundations generally, Arnove put it this way, "foundations...have a corrosive influence on a democratic society; they represent relatively unregulated and unaccountable concentrations of power and wealth which buy talent, promote causes, and in effect, establish an agenda of what merits society's attention...a system which...has worked against the interests of minorities, the working class, and Third World peoples."[15] Venture philanthropy continues the colonial legacy of cultural imperialism described by Arnove through a project of "civilizing the savages." In its updated form fostered by VP, "civilizing the savages" means the imposition of market discipline and corporeal discipline in the form of uniforms, heavy student discipline, standardized testing, and standardization of curriculum that is posited as universally valuable and that stands against the "cultures of pathology" attributed to nonwhite students and particularly African Americans and Latinos.[16] In this view, cultural difference needs to be registered in order to overcome it as an obstacle to the inclusive promise of corporate culture that is positioned as the reward for test-based and standardized educational "achievement."

Venture philanthropies continue the rightist project of traditional philanthropies, but they do so while also aggressively redefining the social contract in ways that gut the sense of public service and public obligation from the institutions they aim to influence. Venture philanthropies participate in both the cultural politics of neoliberalism and the class warfare it supports.

## FRIENDS WITH BENEFITS

In his book *Against Schooling: For an Education that Matters,* Stanley Aronowitz, bluntly writes a seldom spoken truth about the motivation for Microsoft's giving and for the Bill and Melinda Gates Foundation. He highlights that there are tremendous financial (especially tax) benefits for super-rich individuals such as Bill Gates to give away large portions of their fortunes to and through nonprofit organizations.[17] Of course, there are other significant self-interested motivations, including the public relations benefits of promoting oneself as "socially responsible," particularly as Microsoft sought to flout antitrust regulations and use its vast resources to pursue monopolistic consolidations in the software industry. (Attorney Joel Klein aggressively prosecuted Microsoft's monopolistic acts, went on to become head of New York City Public Schools and received millions from the Gates Foundation.)

Philanthropy generally and educational philanthropy more specifically have long histories of functioning both as financial benefits for economic elites and ways of countering negative public perceptions and diffusing class antagonism. In an early public relations stunt to counter immense public antipathy, John D. Rockefeller was famously displayed in depression-era newsreels handing out dimes to the indigent. Henry Ford–built schools, worried not only about his public image but also about the dangerous thoughts his future workers might have on the assembly line. Left-wing critics of philanthropy, often coming from a Gramscian perspective,[18] have emphasized the crucial role played by philanthropy in cementing hegemony by producing consent for conservative economic arrangements and educating citizens to comprehend civil society in ways compatible with ruling class interests. One such Gramscian, Joan Roelofs, points to Marx's recognition in Volume III of *Capital* of how the ruling class is able to stabilize and extend its control by assimilating the intellectuals of other classes.[19]

Andrew Carnegie's exhortation in *The Gospel of Wealth* for the captains of industry to spend their surplus wealth for the benefit of the public appears less than thoroughly virtuous or generous from the critical tradition. The critical tradition recognizes that Carnegie first had to employ vast violence to amass surplus wealth through labor exploitation, wage suppression, attacking unions, and working workers to disablement and death while using political clout to minimize public regulations that would impede wealth accumulation. Carnegie's surplus wealth came first at the expense of the public and through the most aggressive means to exclude the public from participating in the processes of wealth creation. Such exclusions required not only the state-supported economic monopoly over property but also a political monopoly to maintain the rules of the game. In sum, for the critics of scientific philanthropy, the cultural project involved assimilating the intellectuals of subordinated classes and groups into the dominant institutions, creating new dominant educational institutions (like schools, libraries, and museums), and instituting new mechanisms to produce knowledge in ways that reproduce social hierarchies. The "gift" of philanthropy in the age of scientific philanthropy was predicated upon the state-coordinated capitalist plunder of the labor and lives of the working class.

## THE CIRCUIT OF PRIVATIZATION

In the early twentieth century, tax laws strongly promoted foundations. Such laws were the result of deliberate public policy decisions. The United States sustained and supported foundations by "permitting the creation of general-purpose perpetuities dedicated to serving an indefinite class of beneficiaries (through changes in state trust law) and by creating tax deductibility for donors (in both state and federal tax law)."[20] It is remarkable how little of the literature on either foundations or VP generally address this fundamental fact that foundations serve as mechanisms to minimize the taxation of wealth and, as such, largely serve as an entitlement for the richest citizens. While roughly one-half of wealth given to foundations comes in the form of small donations, the real financial benefits go principally to the big givers at the top of the economy who are able to reduce their tax burdens significantly. For every $10 given by the Gates Foundation, $4 is lost from the public wealth in taxes.[21] The philanthropist would otherwise give this money to the public in the form of taxes.

By giving to the foundation, particularly to the foundation that the philanthropist controls, the philanthropist essentially evades the bulk of public control over the use of tax revenue. This means that first, VP in education exists only through public financial subsidy; second, the forgone public tax revenue needs to be understood as being effectively, through the design of public policy, redistributed to the private controllers of the foundation; third, the foundation, which is almost always controlled or directed in its mission by economic elites, uses this public wealth for privately determined purposes; fourth, these purposes tend to align with the material interests and ideological perspectives of private elite power.[22]

In the case of VP in education, this public subsidy for private control of educational policy, practice, and administration can be seen operating in three domains: economic, cultural, and political. Economically, venture philanthropists in education pursue a multifaceted neoliberal agenda of school privatization and deregulation— these involve running schools for profit, entering into lucrative real estate deals related to charter schools, and getting rid of unions that could threaten these plans. Culturally, venture philanthropists pursue a largely rightist form of educational policy and practice, emphasizing test-based accountability and expanded corporate school models such as teacher bonus pay for test scores,[23] school commercialism, corporate forms of school administration, and conservative anti-critical curricula and school models. Politically, venture philanthropists in education actively lobby legislatures and districts to establish a national network of charter franchises, expand charter schools locally, implement vouchers, put in place education tax credits that function like vouchers, and allow for remaking teacher education in ways that remove critical and intellectual content. The coordinated goal of all of these efforts by venture philanthropists is to privatize the public school system by "leveraging" the private foundation resources. In effect, then, the public pays to have its own educational system increasingly directed, controlled, dismantled, and owned by private interests.

For a venture philanthropist such as Gates, Walton, Broad, Fisher, Dell, Milken or others, the attraction is not only tax savings, but the possibility of retaining control over the use of tax money that would otherwise go into the public wealth. It is not a coincidence that the central mission of preparing workers for their corporations is high on the agenda of the venture philanthropists, nor is it a coincidence that educational visions organized by broad-based agenda of challenging social hierarchy would be largely excluded from the perspectives they support. As people who are extraordinarily successful in business, it is not surprising that they would believe that they know best what to do with money in a field such as medicine or education that they may know little or nothing about otherwise. However, such belief hinges upon the baseless assumption that all of these fields are like business or are indistinguishable from business. One of the primary casualties of such thinking is the inability to distinguish private from public goods. It is perhaps more accurate to say that the venture philanthropists participate in redefining the public sphere as a private sphere. A prime liability of such redefinition is that society becomes increasingly understood as a collection of atomic individuals responsible only for themselves. Part of what public schooling represents is the care that each citizen has for the well-being of not just herself and her own child but for everyone else in the society, as well as for the society as a whole. Although celebrated in libertarian literature (the books of Ayn Rand, for example) or popular film (the action hero), the radically autonomous individual is an impossibility in that individuals are always situated in social contexts and cannot evade political, economic, cultural relations, and exchanges. The question becomes what kinds of such relations benefit both the individual and the society as a whole.

For venture philanthropists who understand the individual and the social through neoliberal ideology, every individual is to become an entrepreneur—that is, a business person in pursuit of private gain, always looking to start new enterprises and moving on to the next hot opportunity. The teacher must hustle grants from foundations and test scores for bonus pay; the student must hustle grades to cash them in for further education and later cash; and everyone must hustle against each other. For the entrepreneurial educator and student, image becomes increasingly important. If one shifts the register from the market economy vision of the entrepreneur to the democratic political conception of the citizen, it becomes difficult to see how values of collaboration, deliberation, dialogue, and collective purpose have a role. Rick Cohen puts it well and succinctly,

> Privatizing public education transforms the educational environment from one that builds a sense of collective purpose and nurtures democratic ideals to one that emphasizes individual choice and

makes education a commodity to be produced and consumed in the marketplace.[24]

The social costs of the VP approach to schooling include a loss of the ability to grasp the limitations of applying the business metaphor to education but, more broadly, the limitations of an economy and ecology premised on unlimited economic growth.

In what follows, I will illustrate the privatization agenda of the venture philanthropists in K–12 education by focusing on three central privatization projects: charter schools, vouchers, and tax credits. The venture philanthropists have several other key strategies to further educational privatization, such as funding educational scholarships for private school attendance, political lobbying, and funding research. All of these privatization strategies need to be understood together in order to comprehend the broader privatization project, and these need to be situated in terms of the broader neoliberal project of which VP is symptomatic. My concern is twofold: 1) the public implications of the privatization agenda pursued by VP; 2) how the privatization agenda redefines not only the public sphere but also nonmarket values such as generosity and the gift.

## THE PRIVATIZATION AGENDA OF VENTURE PHILANTHROPY

Studies of VP's promotion of vouchers tend to look at vouchers together with other privatization projects. In a very thorough study critical of privatization, Rick Cohen, writing for the National Committee for Responsive Philanthropy, evaluates a "choice" agenda comprised of vouchers and tax credits. In a study favorable to privatization, Bryan Hassel and Amy Way also look at "choice" promotion projects, but they consider vouchers together with charter schools. There are problems with accepting the framing and terminology of "choice" in studying educational privatization despite the fact that it has won acceptance across the political spectrum. "Choice" is an ideologically loaded term carrying a positive valence. To be critical of "choice" can only position one as an enemy of individual freedom. The language of "choice" naturalizes educational provision through the market metaphors of shopping, treating students as consumers and teachers as providers of a private service. Of course, individuals do not all have the same power in the marketplace, and hence the monetary and non-monetary resources that individuals bring with them determine "choices" in markets. This highlights one of the central dangers of the venture philanthropists' attempts to make education into a market. It stands to worsen rather than to ameliorate the ways that public services are beholden to the wealth of citizens. Instead, one must retain universal public schooling as an ideal referent in discussing privatization, consider the elements and strategies of the privatization proposals, and not fall prey to the market metaphors like "choice" that have shifted the terms of educational policy debate.

The National Center for Responsive Philanthropy studied 132 organizations specifically dedicated to vouchers and or tax credits between 2002 and 2006, "think tanks, advocacy organizations, parent organizations, and educational scholarship organizations that provided research and promoted support" for vouchers or tax credits. They found that, in these four years, 104 of these 132 organizations were receiving grants from 1,212 different foundations. According to Hassel and Way in the context of educational grant giving, Walton Family Foundation, Gates Foundation and Casey Foundation stand out among foundations giving major portions of their annual funding to privatization.

A single foundation, The Walton Family Foundation, "dwarfed all other foundation funders of this movement"[25] when defined through vouchers, scholarships, and tax credits. In 2005 alone, Walton gave $25 million to organizations promoting vouchers and tax credits out of over $86 million given by 495 different foundations.[26] Walton gives $25–30 million a year to voucher and tax credit organizations. Walton makes a clear distinction between "public charter schools," and what is meant by "school choice." Walton's focus is to use its resources for "low-income students to choose and attend private schools."[27] Walton describes its work as comprised of the following: building support for public policies that favor privatized schooling; strengthening the public funding of scholarship programs to pay for private schooling; information dissemination to parents about public, charter, and private school options; and evaluating the performance of privatization programs.[28] These initiatives use the private wealth accumulated through the Wal-mart Corporation to influence and encourage students to use and favor privatized over public schooling. In addition to eroding the base of students for public schools, this erodes the political will to strengthen and support quality universal public schooling. Part of what makes this undermining of public schooling particularly egregious for the inheritors of the Wal-mart fortune is that the Wal-mart wealth comes from aggressive anti-public and antihuman union-busting. Inexpensive products on the shelf are the result of low wages, no benefits and other bad labor conditions in both North America and in the labor forces in the nations engaged in the global race to the bottom. What should not be missed here is that the Walton family seeks to apply radically unequal private sector practices to the public sector. While Wal-mart has been particularly successful at expanding the availability of cheap consumer goods, they have also been widely credited with decimating small businesses and local communities, transforming the workforce to super-exploited labor, and spending vast sums to prevent workers from organizing to fight these conditions. Wal-mart has made billions by diminishing the wealth and standards of living of citizens.[29] This impoverishment weakens local tax bases and in effect defunds schools and other public sector services. Waltons then take the profit gained and use it to further erode local public schooling. This should be understood as a concerted neoliberal effort to destroy the public while strengthening private power.

The range of financial support Walton gives to privatization outfits includes organizations such as its number one recipient: the Children's Scholarship Fund. In 1998, the Children's Scholarship Fund was started with matching $50 million grants from Walton and financier Theodore Forstmann. Walton added more than $94 million between 2000 and 2005 alone. Walton commonly gives multi-year seven and eight figure grants to educational privatization organizations.[30] As of 2005, 23,000 children were funded for private schools.[31]

As of 2006, seven states offer tax credits or tax deductions for charitable giving to fund students to attend private schools rather than public schools. In Pennsylvania, 2200 corporations took advantage of the tax credit program and gave scholarships to 33,000 students in nearly every county of the state. Yet, as of 2005, 70 percent of tax credit funding encouraged student attendance in private schools rather than in public schools.[32] The public is subsidizing the private sector to weaken the public sector. These tax credit-supported scholarship projects to expand private schooling and weaken the base of support for public schooling are being aggressively supported by the venture philanthropies.

The list of recipients of more than $100,000 over five years includes the major pro-privatization think tanks and political lobbying organizations including the already quite well endowed Alliance for School Choice, Hoover Institution, the Milton and Rose Friedman Foundation, Thomas B. Fordham Institute, Manhattan Institute, Heartland Institute, American Enterprise Institute for Public Policy Research, Hudson Institute Inc. High on the list in terms of money received are the Hispanic Council for Reform and Educational Options and Black Alliance for Educational Options Inc. Heavy funding for pro-privatization minority organizations represents an important effort by privatization strategists to align populations historically injured by public schooling. Traditionally, public schooling has been beholden to the racialized dimensions of economic inequality and a culture of white supremacy that organizes educational policy and practice. As Thomas Pedroni details, right-wing think tanks and foundations bring minority parents into political coalitions albeit in subordinate positions.[33]

The Center for Responsive Philanthropy report describes the levels of grant funding by Walton on K–12 reform as "astounding... by any calculation," that they have been "redefining the meaning of philanthropy in ways unparalleled" and that Walton grants "are of the size and focus to create and sustain powerful and effective organizations for K–2 education reform."[34]

Openly critical of educational privatization, Cohen's report argues that vouchers weaken the public commitment to public education by allowing parents to opt out of public education: this weakens the base of support for public education. The proponents of privatization wholeheartedly agree that subsidizing scholarships, vouchers, and tax credits will indeed get the public to opt out of public education. Where they disagree is over the matter of whether public schooling should be, in the words of Cohen, "one of the deliverables" of government.

As the opening of this chapter illustrates, venture philanthropists and other advocates of privatization display an animosity to public education and a desire to destroy it rather than strengthen it. On the other hand, they hold a deep faith that markets can do anything better than the public sector can, and they believe that consumer choice works best as a form of public policy. Tell that to people who made rational consumer choices to buy subprime mortgages, SUVs, fast food lunches, and cigarettes.

Regarding subprime mortgages, SUVs, and fast-food lunches respectively, consumer choice frequently becomes destructive despite the best intentions of the consumer when the commodity in question can be transformed with the vicissitudes of the market, or when the commodity relies upon another commodity (oil) that is subject to the vicissitudes of the market, or when the commodity is a convenient choice in the short run but a destructive choice in the long run (burgers and oil). The promise of the neoliberal market celebrated by VP is the promise of continual economic growth for continually rising prosperity. Markets do not continually grow. They also shrink, and frequently economic growth is unequally distributed. So when economic growth is applied as a metaphor to a public service like education, the public sector inherits the instability of markets. As even the popular press has come to admit, ever expanding economic growth is both ecologically unsustainable and inevitably makes economic losers.[35]

One of the grand contradictions of this attempt to turn public schools into a market is that the means to do so are being largely pushed using nonmarket means—that is, through nonprofit entities. When the largest for-profit manager of public schools, the Edison Schools, was being publicly debated in Witchita, Kansas, the local newspaper published an editorial that criticized the basic premise of EMOs such as Edison. It said that if these for-profits EMOs rely on grants for their operations, then, they look less like business and more like "panhandling."[36] It is crucial to recognize that the push for privatization through the use of nonprofit and state institutions represents a right-wing movement. This is a movement that can be understood and stopped with adequate public will.

Of foundations giving more than $500,000 a year to vouchers and tax credits (for 2005) Walton ($25,343,778) tops the list by far. However, the other leading givers include the right-wing Lynde and Harry Bradley foundation ($6,358,000), Sarah Scaife Foundation ($3,895,000), Bill and Melinda Gates Foundation ($2,624,387), John Templeton Foundation ($2,560,014), Herrick Foundation (2,370,000), Annenberg Foundation (2,100,000) among many others.[37] These big givers alone gave over $65, 526,652 to vouchers and tax credits in 2005—a year considered by Cohen to be a watershed moment in the pro-voucher movement.[38] However, the quantity of financial giving to vouchers and tax credits does not convey the strategic significance of the privatization movement.

Pro-voucher forces are comprised of organizations dedicated strictly to voucher promotion and organizations that are multi-issue right-wing organizations. The latter use their broader advocacy power to further the voucher agenda. Examples of these include right-wing think tanks like the Heritage Foundation, Hoover Institution, American Enterprise Institute, Manhattan Institute for Policy Research Inc., Cato Institute among others. This list also represents five of the top ten voucher and tax credit recipients of general support grants in 2005.[39] General operating support grants allows these organizations to have long-term stability and develop comfortably long-term strategies to achieve the goals of furthering public sector privatization through disseminating research, lobbying policymakers and scholars, promoting their ideas in the popular press, funding local and national initiatives, producing conferences, and so on. Liberal and left foundations do not receive nearly the level of general operating support of the right-wing foundations. The top five of the foundations that give general operating support to these think tanks are (for 2005): Bradley ($3,045,000), Scaife ($2,545,000), Annenberg (2,000,000), Hume (1,955,262), and Samuel Roberts Noble (1,600,000).[40]

As Cohen points out, the effectiveness of the voucher movement is not simply a matter of "adding up totals" but comprises a combination of

> the kinds of grants (general operating vs. project or program-restricted), the availability of grant dollars for multiple years (addressing nonprofit sustainability), the numbers of grants (reaching more groups), and the geographic dispersion of grant recipients (toward constructing an infrastructure of organizations).[41]

Cohen contends that, all told, the picture of right-wing grant giving for vouchers and tax credits demonstrates a "collective commitment to intelligent grant making that serves the needs of its movement partners."[42]

While the political right has dwarfed the liberal involvement with VP, the election of Barack Obama in the U.S. presidency, his reinstallation of much of the Clinton administration and, in particular, his appointment of former Chicago education "CEO" Arne Duncan as U.S. Secretary of Education has meant that organizations such as Center for American Progress (founded by Bill Clinton's chief of staff John Podesta) and EPI which are lined up with several aspects of the VP agenda have gained power and prominence. Most notably, while the Obama administration rejects vouchers, it nonetheless emphatically champions charter school expansion, numerical test-oriented definitions of student achievement, "data driven" policy-making, and early indications of threats to both teacher education and teachers unions. Most significantly, Obama appears to accept the VP idea of "leverage" framing of public schools needing to be forced to compete and "choice" as a mechanism to do so. Obama's embrace of charter schooling means that one of the central VP projects is at the center of the federal agenda.

Venture philanthropists in education have a long-term, top-down strategic vision but they also understand the value of empowering and funding, with minimal constraints, bottom-up grassroots organizing for vouchers and tax credits at the local level. In fact, private non-corporate funders of privatization give much higher percentages of their assets annually than is either the norm in the foundation world or is the legal minimum of five percent. One hundred and fifty-four funders gave more than ten percent of their assets.[43]

Political activism plays a significant role in the strategy to achieve vouchers, in addition to the nonprofit 501(c)(3). Political organizations 501(c)(4) are tax exempt and appear similar to 501(c)(3)s, but do more political lobbying and other direct political influence. Giving to a 501(c)(4) does not offer tax deductions. However, 501(c)(3)s and 501(c)(4)s are often closely related, and many 3s create subsidiary 4s to do their more overtly political work. An example of this would be the neoliberal Alliance for School Choice that is both 501(c)(3) and 501(c)(4). They played a crucial role in lobbying for the radically privatized rebuilding of the New Orleans schools following hurricane Katrina.[44] Examples of 501(c)(4)s include Parents for Choice in Education (Utah), Advocates for School Choice (AZ), DC Parents for School Choice (DC), Floridians for School Choice Legislation (FL), Minnesotans for School Choice (MN), Parents for School Choice (NY), Parents for School Choice (MO), American Civil Rights Coalition (CA), Americans for Limited Government (IL), Institute for Educational Reform (VA), Parents Alliance for Choice in Education (CO), and United New Yorkers for Choice in Education (NY).[45]

Other nonprofits like 527s exist specifically to lobby for political elections while Political Action Committees raise money for a specific candidate. Unsurprisingly, voucher and tax credit- based campaign-giving goes overwhelmingly to Republican candidates.[46] While a 2002 Supreme Court decision set the stage for the further expansion of vouchers at the local level, in 2005, Katrina was used to justify a temporary national voucher scheme, and, in 2008, Louisiana passed a statewide voucher law, the first of its kind.

The creation of the first voucher scheme in Milwaukee in 1990 was a crucial step in the promotion and expansion of vouchers by privatization advocates. By the 2003–2004 year, 14,000 children were involved.[47] As with other privatization initiatives, the first project is strategically crucial for advocates to begin to churn out favorable research and comparisons with public schools and also use the project as a launching pad for replication even if the evidence is largely negative, inconclusive, or simply not yet available. Without the existence of any voucher projects, it was difficult for privatizers to justify introducing any, not to mention securing, funding.

As Cohen writes, it is not a coincidence that the political will and public sentiment for the expansion of privatization grows in the United States.

> Support for school vouchers, school choice and education tax credits does not exist simply because of the hopes, aspirations, disappointments, and frustrations of school children's parents. Their awareness of school privatization options is supported by an array of advocacy institutions, think tanks and scholarship organizations that function to create a consciousness that there are alternatives to traditional public schools. The Cato Institute, the Heritage Foundation, the Heartland Institute and many others monitor and promote news about polls, bills, and programs, placing articles and op-eds in local and regional newspapers around the nation to maintain and increase public awareness and support for school privatization.[48]

What should be clear is that while the networks of right-wing think tanks detailed by Cohen play a crucial role in educating the public toward the neoliberal school agenda, these educational projects are part of a broader constellation of what Henry Giroux calls "public pedagogy." The crucial language of "public pedagogy" in this instance highlights that the educational projects of the political right educate the public into consent or even desire for privatization. What the language of "public pedagogy" also illuminates in this case is that these private forces are producers of meanings, identifications, and ideologies in the public sphere, often in ways that result in the weakening of public institutions. Giroux writes,

> Under neoliberalism, dominant public pedagogy with its narrow and imposed schemes of classification and limited modes of

identification use the educational force of the culture to negate the basic conditions for political agency...Politics often begins when it becomes possible to make power visible, to challenge the ideological circuitry of hegemonic knowledge, and to recognize that "political subversion presupposes cognitive subversion, a conversion of the vision of the world." But another element of politics focuses on where politics happens, how proliferating sites of pedagogy bring into being new forms of resistance, raise new questions, and necessitate alternative visions regarding autonomy and the possibility of democracy itself.[49]

Giroux's point highlights the political stakes in the pro-privatization education projects of the political right as well as the possibilities for those committed to public democracy to interpret such pedagogies and produce democratic pedagogies. This insight is important as well for illuminating the limitations of the dominant liberal policy "wonk" responses to the activities of the right-wing think tanks. Most liberal critics of the privatization movement and the charter movement challenge the numerical claims of the privatization propagandists while failing to call into question the assumed goal of educational efficacy and the ways that the struggles waged for educational policy are interwoven with class and cultural struggles.

Hassel and Way define "choice" through both "private choice" vouchers and tax credits and "public choice" public charter schools. They contend that the Gates and Walton foundations together comprise 87 percent of "choice related funding."[50] In part, this is significant because it indicates an extremely concentrated privatization effort by a small number of people with tremendous resources. As Frederick Hess of the American Enterprise Institute recognizes, this raises serious questions about the accountability of those few capable of directing so many resources that can leverage educational policy and lead reform. Unfortunately, Hess and the majority of contributors to his pro-privatization edited collection *With the Best of Intentions* presume that despite these concerns, the government should under no circumstances more closely regulate the uses of this publicly subsidized private money as it influences the public. Consistent with the neoliberal antipathy to public schooling, these contributors largely believe that the "best intentions" of the public sector cannot be relied upon, and instead the magic of the market should be relied upon. Hess goes so far as to invoke Tocqueville to argue that public participation is really only private participation.[51]

## CHARTERS

Charter school promotion is a central aspect of the VP project of educational privatization. Proponents of "school choice" frequently make a distinction between "private school choice" (vouchers, scholarships for private schooling, and tax credits) and "public school choice" (charter schools).[52] This division obscures the ways that charter schools further the overall privatization and corporatization of public schooling in a number of ways.

Charter schooling has to be understood in relation to the broader neoliberal agenda that is comprised of not just privatization, but deregulation and what it enables such as anti-unionism and contracting. Simply in terms of privatization, charter schools allow for-private for-profit companies to manage schools. For example, half of the Edison Schools, the largest for-profit company managing schools, are charter schools. As of 2008, "charter schools continue to account for approximately 85 percent of all EMO-managed schools."[53] The last decade has seen a steady rise in the number of EMO-managed charter schools while the number of EMO-managed district schools has not increased significantly in the same time, and the number slightly declined in 2008.[54] This suggests that charter schools are at the forefront of the educational privatization movement. Moreover, the vast majority of EMO-managed schools, 84.4 percent, are managed by large EMOs[55] that tend to replicate the same school model. This suggests that the original impetus for charter schools to offer diverse school models and experimentation is being utterly reversed by the involvement of EMOs. In places where for-profits are prohibited frequently, nonprofit companies subcontract with for-profit companies to manage schools. That is, so-called charter management organizations (CMOs) subcontract to EMOs. As well, charters do not receive many district supports and further contract out with for-profit companies for services that range from education to transportation and food.

Teachers in charter schools have been prohibited from forming unions with school districts. Anti-unionism has been central to the justifications for charters, although this is most frequently described as being about the virtue of freedom from bureaucracy. As a consequence, most charter schools do not have unions. For example, in the Chicago Public Schools, no charter schools have union representation. This erodes the power of teachers to not only control their work conditions, but also weakens the power of teachers unions in districts and nationally. The erosion of teachers' power over their job security, work conditions, and the content of what they teach has effectively transferred educational governance power to city, state, and federal government officials who have been largely won over by the conservative educational agenda that is favored by the privatization movement. Chicago's mayoral takeover of the public schools allowed for the school reform to be written by the Commercial Club of Chicago, and this has involved a vast privatization and deregulation agenda that includes antiunionism, contracting out to EMOs, smashing elected community school councils, and generally seeking to create a "market" in public schools by creating charter schools and spin-off forms of charters to circumvent state limits.

In Chicago, Arne Duncan on behalf of Mayor Daley and the Commercial Club of Chicago (which wrote the plan) implemented the Renaissance 2010 plan that is deeply aligned with the VP agenda. In the spring of 2008, the funding arm of Renaissance 2010 held an event mc'ed by Duncan called "Free to Choose Free to Succeed: The New Market in Public Education." Venture philanthropists including Gates, Charter School Growth Fund (CSGF), and New Schools Venture Fund (NSVF) as well as the Fordham Foundation had a prominent presence at an event that involved open discussions of how to create national charter franchises, and how to overcome union and state legislative opposition to unchecked expansion of charters. Duncan himself employed the VP rhetoric, speaking of the Chicago Public Schools as his "portfolio" and the need to "leverage" change in the public system through competition and choice.[56]

Charter schooling has garnered support across the political spectrum because of the early promise of innovation, experimentation with new school models, choice, competition, decreases in bureaucratic constraints that inhibit innovation and experimentation, and "effective delivery" of education. Venture philanthropy, in particular, has pushed for the replication of charter schools to create "charter franchises," and it has aggressively pushed to expand the number of charter schools in districts in order to "leverage" influence in those districts in ways favorable to the agendas of charter and privatization promoters. The NSVF and the CSGF are two prominent venture philanthropies aggressively seeking to promote both "charter franchises" and "leverage" on the public school systems while treating charters like businesses, school operators as entrepreneurs, and groups of schools as "portfolios." These charter-support organizations share the vision of the right-wing think tanks in that they openly discuss the ideal of creating a "market" out of the public schools locally and nationally. They idealize competition and choice and talk about how the creation of charters will force the public schools to compete with the charters for resources and students, and that this will, "just like in business," force schools to increase quality or "go out of business." The charter-support organizations, including NSVF and CSGF,[57] are heavily funded by Gates, Walton, Broad, Dell, and Fisher foundations in particular as well as by other foundations.

The VP emphasis on "replicating" school models that are deemed successful and "franchising" charters results in the homogenizing of school models. Hence, the early justifications behind the charter movement of "innovation" and "experimentation" are undermined by the VP insistence on "replication" and "scaling up" which are terms borrowed from venture capital. This sentiment was expressed to me by one accountant for charter schools in attendance at the "Free to Choose, Free to Succeed: The New Market in Public Schools" event put on by the Renaissance Schools Fund in Chicago. He lamented that the reason he got involved with charter schooling was the promise of innovation and experimentation that appeared to be getting sacrificed by the application of the "franchising" ideal dominating the charter movement under pressure from VP. What is more, there is little evidence that the push to "replicate" and "scale up" particular school models has to do with any coherent definition of quality. For example, in Chicago, Perspectives Charter School has been encouraged to replicate despite evidence of mediocre to poor test score achievement. Such replication flouts state limits on charter schools by creating "satellite" campuses and thereby circumventing legislation designed to protect public schools from charters. In some cases, replication is justified through test scores; in other cases through other criteria. One panel at the "Free to Choose, Free to Succeed: The New Market in Public Education" event focused on replicating successful school models. There was a principal from one Chicago lottery enrollment school, that is K-6, that is in a rich neighborhood, and that receives a massive amount of grant money. The school was "replicated" through the assistance of a company out of New York called "Replications, Incorporated." The new principal of the replicated school spoke of how her school was community enrollment, K-3, and how it was receiving no

outside money and hence was financially strapped. The two principals and the "Replications, Incorporated" consultant agreed at the
end of the panel that in fact "replication" is a myth because inevitably the school being created is never going to have the same conditions, resources, and staff as the model school. But, the three
emphasized, while replication is a myth, it is a myth that "works"
to get schools created. Of course, what went unsaid was that in
Chicago these schools are being created after Duncan's destructive
"turnaround" approach that has shut down entire schools and/or
fired every last employee right down to the janitorial staff.[57]

In fact, the 2009 killing of Derrion Albert in Chicago, which
was videotaped and widely viewed on the Internet, symptomatizes
the destructive effects of these efforts. Duncan closed the Altgeld
Gardens high school to transform it into the Carver Military
Academy, forcing students from Altgeld to attend Fenger Academy.
Fenger had just been "reconstituted" with every employee replaced
but the same students were in place. The Altgeld students were
added to the high school and the predicted and predictable results
were gang violence. Both Duncan and Huberman were warned by
community members that these policies of closing and reopening
experimental schools without attention to gang and neighborhood
affiliation would result in students being forced to cross gang and
neighborhood lines and would result in both violence and dropping
out as a strategy by students to avoid violence. By adding the Altgeld
students with no continuing teachers, administrators, and counsel-
ors familiar with the community history and dynamics, this
approach was a recipe for disaster. Duncan's decision to "turnaround" Fenger in this way followed at least three failed attempts at
reform and the infusion of millions of dollars in Gates Foundation
money that was used to implement a "turnaround curriculum,"
which caused no improvement in academic achievement traditionally defined. Gates withdrew funding at the start of 2009 prior to
Duncan's decision to do the mass firing. Following the killing,
Duncan was interviewed about how his policies had set the stage for
the Fenger killing on National Public Radio. Duncan flatly denied
any responsibility.

Defined through "choice," the two venture philanthropies of
Walton and Gates completely dominate foundation-giving in total.[58]
Relative to total foundation-giving specifically to charter schools,
the Gates Foundation is by an overwhelming level the dominant
funder. As of 2002, 94 percent of Gates "choice related funding"

went to "public school choice" including charter schools, small schools, and alternative schools.[59] When Hassel and Way looked at charter school funding as of 2002, they found—after removing Gates and Walton—only eight percent of foundation giving for "choice" went to charter schooling while 81 percent went to private school grants.

The venture philanthropies and other foundations have "supply side" strategies that are designed to increase the number of schools outside of public school districts. So they fund networks of charter schools such as Aspire and Kipp. "What they have in common is an insistence that member schools meet certain criteria and maintain certain brand standards. Many funders hope these brand-name organizations will foster greater quality on average than stand-alone schools and achieve economies of scale in their operations."[60] Perhaps the clearest evidence of the incoherence of the economies of the scales argument can be cited as the fate of the largest for-profit manager of charters, The Edison Schools. The Edison Schools project was justified on the basis of eventually achieving economies of scale by becoming the largest quasi-district in the United States and then being able to cut costs. This did not work for Edison, and the necessity of drawing profit out remains a distinct disadvantage that the for-profits have in this regard. Many nonprofits that are subcontracting with for-profits face the same problem. Furthermore, public systems already benefit from "economies of scale," and it is unclear why fragmenting them with charters would help rather than hinder any such benefits.

The venture philanthropies also promote "demand side" strategies. They fund "choice" and support organizations with specific purposes such as Building Excellent Schools, New Leaders for New Schools (fostering specific kinds of educational leadership), Civic Builders (supporting real estate development and the financing of facilities), National Council of La Raza and Charter School Development Initiative (targeting Latino/as). These charter advocacy organizations both provide financial and organizational support for charter schools, and they advocate for charter schools.[61] With so much money coming from so few venture philanthropists pushing privatization, these organizations give the false appearance of bottom-up support spontaneously developing for privatization— another name for such fake grassroots support is "astroturf."

"Demand side" strategy includes scholarship funds for private schooling such as the aforementioned Children's Scholarship Fund funded by Walton, information-oriented organizations like GreatSchools.net, or magazines like *Education Next*, intended to lead parents to privatized school alternatives and grassroots-organizing. Examples of grassroots-organizing include the Black Alliance for Education Options that has a national organization and has received significant start-up funding from Walton as well as ongoing support from Bradley and other foundations. So-called "demand-side" organizing includes advocacy organizations influencing policy-makers and legislators and supporting pro-privatization research. Prominent examples include Harvard's Program on Education Policy and Governance (Paul E. Peterson), Manhattan Institute's Education Research Office (Jay P. Greene), and Charter School Research Project at the University of Washington Center on Reinventing Public Education (Paul T. Hill). These right-wing intellectuals and institutions have a high degree of coordination, appearing in the same publications and at the same conferences, and they receive heavy levels of scholarly support.

Charter-support organizations range from the local and state level (California Charter Schools Association; New Visions for New Schools (NY)). These are but two of numerous organizations throughout the United States including Alliance for School Choice, Black Alliance for Educational Options, Center for Education Reform, Charter School Leadership Council, and so on. These support and advocacy organizations help charter schools in multiple ways from creating relationships with vendors and insurers, facilitating real estate deals to public relations and other influence peddling.[62]

## CONCLUSION

Venture Philanthropy's concentrated financial support for these privatization initiatives and coordinated anti-public movement connecting research, public relations, grassroots- organizing, policy influence, and political lobbying does not merely raise issues of public governance but also the specter of a small number of people being able to destroy the public school system as a public system. What is being accomplished in the name of excellence and school improvement is a largely long-standing right-wing agenda of weakening rather than investing in public schooling, smashing teachers' unions, putting in place for-profit and nonprofit charter schools, vouchers, tax credits, and scholarships.

Public schooling is crucial for a democratic society. Strengthening public schooling involves increasing and equalizing school funding, investing in teacher's work, improving teacher's work conditions, investing in public school physical sites, expanding the range of public school services to increasingly integrate with other public services such as child care, nutrition, extracurricular, social and civic events and other community activities. Strengthening public schooling also involves reinvesting the public school with its role as a site for deliberation and action in a public democracy.

## 3

# FROM TRADITIONAL TO VENTURE PHILANTHROPY

Venture philanthropy differs markedly from prior educational philanthropy, dominant throughout the twentieth century including large donors such as The Carnegie Corporation, The Rockefeller Foundation, and The Ford Foundation. These traditional philanthropic endeavors were defined through a sense of the public obligation of industrialists to give back some of the surplus wealth that they had accumulated. Carnegie's *The Gospel of Wealth* codifies this perspective that its advocates described as "scientific philanthropy."

As critics such as Robert Arnove, Joan Roelofs, and others have argued, the early educational philanthropy played a distinctly conservative cultural role of supporting public institutions in ways compatible with the ideological perspectives and material interests of the captains of industry rather than of the workers of coal, steel, oil, or automotive production. Such labor created the surplus wealth that then went into educational institutions, museums, libraries, and trusts. Public subsidies through tax incentives not only encouraged but financed such public works to be developed and designed by fiscal and cultural elites rather than by the broader public.

Although educational philanthropy played a hegemonic role throughout the twentieth century, it was hardly unified in its approaches and offered funding for a wide variety of initiatives and projects that were not restricted to the conservative side of the political spectrum. There was a distance between the donors and the uses made of the money in education. "Scientific philanthropy," though beholden to the logic of cultural imperialism,[1] was marked by a spirit of public obligation and deeply embedded in a liberal democratic ethos.

Venture philanthropy departs radically from the age of "scientific" industrial philanthropy. Venture philanthropy is modeled on venture capital, in particular, its investments in the technology boom of the early 1990s. VP is consistent with both the upward material distributions of a "new guilded age" and the steady expansion of neoliberal language and rationales in public education, including the increasing centrality of business terms to describe educational reforms and policies: choice, competition, efficiency, accountability, monopoly, turnaround, and failure. Likewise, VP treats giving to public schooling as a "social investment" that, like venture capital, must begin with a business plan, involve quantitative measurement of efficacy, be replicable to be "brought to scale," and ideally will "leverage" public spending in ways compatible with the strategic donor. Grants are referred to as "investments," donors are called "investors," impact is renamed "social return," evaluation becomes "performance measurement," grant-reviewing turns into "due diligence," the grant list is renamed an "investment portfolio," charter networks are referred to as "franchises," to name but some of the remodeling of giving on investment and particularly on venture capital models.

Within the view of VP, donors are framed in private terms as both entrepreneurs and consumers while recipients are represented as investments. One of the most significant aspects of this transformation in educational philanthropy involves the ways that the public and civic purposes of public schooling are redescribed by VP in distinctly private ways. Such a view carries significant implications for a society dedicated to public democratic ideals. This is no small matter in terms of how the public and civic roles of public schooling have become nearly overtaken by the economistic neoliberal perspective that views public schooling as principally a matter of producing workers and consumers for the economy and for global economic competition.[2] Rather than breaking with the neoliberal economic assumptions about education that intensified throughout the Reagan, Bush, Clinton, and Bush years, the Obama administration displays a deep commitment to expanding radically the twin imperatives of neoliberal education in the form of privatization and deregulation: charter schooling (despite a lack of compelling evidence for its boasted successes); implementing Wall Street–style bonuses tied to test scores for teachers and students; pushing urban education projects such as those in Chicago and New Orleans that are tied to public housing privatization and destruction, and that dispossess citizens of their communities while deunionizing school districts.[3] The local business groups that coordinate with venture philanthropists push such neoliberal urban education reform initiatives.

In what follows, I first show how VP constitutes a radical break with prior forms of philanthropy in terms of its underlying assumptions and animating language. First, I look at its antecedent "scientific philanthropy" and then at its origins. I then consider the ways that VP expands corporate culture in both the philanthropic and educational domains. I conclude by discussing how VP is not merely a reflection of particular values and interests but also functions pedagogically by producing particular ideas and ideologies about educational obligation in a highly privatized way. Of chief interest is the question of why educational philanthropy serves capital interests differently in the age of scientific philanthropy as opposed to in the age of VP.

## THE USES OF "SCIENTIFIC PHILANTHROPY"

A leading scholar on philanthropy, Stanley Katz specifies the particularly public understanding held by early twentieth-century leaders of "scientific philanthropy." ". . . I want to use the term "philanthropy" in the special sense originated by Carnegie and the senior Rockefeller: as the self-conscious donation of truly large sums of private wealth to do public good by addressing the causes (and also manifestations) of social problems of all kinds."[4] It should be said from the outset here that for critics on the left, the causes of social problems are the social structures and systems that facilitate the vast amassing of wealth by few at the expense of many. The projects of Carnegie and Rockefeller were defined through the public interest and appear on one level to be concerned with redistributive efforts toward ameliorating inequalities in wealth and income as well intertwined cultural inequalities such as unequal access to education. However, as Katz among others points out, the scientific philanthropists were deeply conservative and understood their giving as having a practical use for themselves and others of their class.

> They believed that the private sector needed to step up to enhance the public welfare, both to relieve the political pressures of popular unrest and to reduce the chances that the state (especially the federal government) would rise to that task.[5]

Scientific philanthropists sought not simply to use private money for public gain but to serve ruling class interests in a number of ways. These conservative intents include delegitimating socialist politics and movements, establishing institutions that directly serve elites, assuring social reform rather than radical structural change, and creating social networks to secure the status of elites.[6] Additionally, foundations have supported social programs such as social security to assuage depression-era labor unrest, and they worked to support tax laws that prohibited giving to political parties, support for civil rights and minority education projects in part to diffuse minority interest and support for radical movements; and foundations supported "democracy promotion" projects overseas that would increase the likelihood of political economic formations tending toward liberal capitalism rather than socialism or communism.[7] Carnegie exhorted the super-rich to found universities, and many did, such as Cornell, Stanford, Johns Hopkins, and Rockefeller's University of Chicago. Stanford and Chicago have been hotbeds of neoliberal thought: University of Chicago arguably was the birthplace of neoliberalism under Milton Friedman (some would claim London School of Economics and Hayek) and leaders of the push to privatize public schooling are associated with the Hoover Institution housed at Stanford. This is not to say that there has been no progressive or radical political thought coming out of these universities but rather a recognition of the centrality of foundations to the early formation of educational policies that have left a conservative legacy.

As Roelofs argues, nearly all public education reform of the twentieth century has its origins in philanthropy.

> Nearly all reforms in public (as well as private) education originated with foundations. The course credit system and centrally administered college entrance examinations came about as a requirement for the college teacher's pension program (now TIAA-CREF) started by the Carnegie Corporation for the Advancement of Teaching. These had a major effect on standardizing high school education throughout the United States, as college admission increasingly dictated curricula. Carnegie later initiated "new math," "Sesame Street," and "service learning." Ford, along with Carnegie, was a major promoter of educational television developers of the Corporation for Public Broadcasting, Headstart, Upward Bound, and alternative schools. In 1967, McGeorge Bundy, President of the Ford Foundation, was appointed by New York City's mayor as chair of a task force to plan for NYC school system decentralization.[8]

The scope of liberal and sometimes even progressive commitments to the public sector emerging from scientific philanthropy neither invalidates the conservative project of undermining radical movements for systemic change and genuine democracy, nor does this history invalidate the reality that the commitment to the public good did result in the strengthening of the deliberative aspects of the public sphere. Both aims can be found explicitly stated in Carnegie's *The Gospel of Wealth*. He writes that the best gift philanthropy can give to a community is a free library, "provided the community will accept and maintain it as a public institution, as much a part of the city property as its public schools, and indeed, an adjunct to these." Carnegie emphasized the value to the public of the free access to knowledge and information, and he understood public knowledge institutions as ameliorative by allowing the poor opportunities for self-advancement.

Carnegie, like the leading educationalist of his day, G. Stanley Hall (who is largely responsible for the late-nineteenth-century field of the study of adolescence), accepts the racially grounded doctrine of recapitulation theory. Recapitulation theory holds that the development of the human being repeats the development of the human race and that, the successful development of the human race toward civilization depends upon youth being forced to undergo the trials of earlier stages of human development. These trials build character in middle-class white boys and prepare them to lead civilization forward.[9] Such movements as scouting and the YMCA typify such early twentieth-century thinking that viewed getting back to primitive nature as a necessary strengthening endeavor to prepare youth for stewardship of civilization. In the view of recapitulation theory, human development follows from primitive nature to animals to lower human to higher humans. Within this schema, white European males are at the top of the upward chain of nature. However, white boys, in particular, need to go back down the chain to get toughened up for the stresses of governing advancing civilization. In the *Gospel of Wealth*, Carnegie celebrates his own impoverished childhood and the character he gained by working as a child laborer in the textile industry. He describes his adult visit to the home of a Sioux Indian chief and makes much of the fact that the lowliest Indian and the chief live in indistinguishable dwellings. For Carnegie, this illustrates the superiority of Euro-American civilization. The difference between the worker's cottage and the millionaire's mansion indicates for him an upward movement toward greater and greater civilization. He argues that capitalism raises everybody's quality of life and that, the amenities of the worst-off in civilized society are superior to the living standards of kings in prior eras. However, competition and the refusal of aristocratic inheritance make possible the forward movement toward greater and greater innovation and civilization. Carnegie extols the virtues of poverty and the valuable lessons bestowed upon child laborers, and also laments the misfortune of the children of the rich who do not benefit from the character-building blessing of destitution. Carnegie sees the Sioux as both stuck in the prior history of the human race and as communistic—communism, Carnegie explains is not progress but regress, bringing humanity back to the life-standards of "primitives." For Carnegie, capitalism produces both wealth and poverty. The dim and stultified aristocrats of old Europe have suffered from the mistake of inheritance. Carnegie exhorts his millionaire contemporaries to be ashamed to die with their wealth. Instead, they ought to give it to the public so that those who can help themselves will do so. Those incapable of helping themselves should be left to the care of the state, he explains.

Central to Carnegie's view of philanthropy is the value of self-help but also the definition of human worth through economic productivity. Carnegie rails against the violence of frivolous giving of charity, suggesting that the nickel given away on the street goes on to do compounded harm to the recipient whose productive energies will be drained by the possibility of unproductive acquisition. Scientific philanthropy for Carnegie must be highly rationalized based on its inspiration for fostering economic productivity. However, it also must contribute to the public good that cannot be strictly reduced to the economic. Indeed, Carnegie has harsh words for the wealthy person who flaunts wealth in conspicuous displays rather than by giving to the public. And Carnegie opposes the giving of vast private inheritances to children, seeing this as a diffusion of productive energies and a corrupting influence. Perhaps, what is most significant in Carnegie is the expansion of a perspective toward wealth found in Benjamin Franklin's autobiography. Franklin taught his readers to view money as having a life of its own and a reproductive capacity. The squandering of wealth was akin to killing productive offspring. Of course, both Franklin's and Carnegie's view of wealth, as being strictly guided by rational utility, typifies the increased rationalization of giving in accord with capitalism.[10] Carnegie's vision for philanthropy deeply displaced a value on dispensing wealth and marked a turn toward the shift from *charity* to *philanthropy*. Bill Gates and other venture philanthropists mark another significant shift in the western understanding of giving.

Bill Gates read Carnegie in preparation for establishing his Bill and Melinda Gates Foundation. There are certain elements of Carnegie's thought that Gates continues, including the rationalization of philanthropy as necessarily fostering "productive" individuals and greasing the inclusion of working people in the ideologies of a corporate-dominated economy that mostly undermines their own interests. However, there are numerous glaring differences between the social visions of Carnegie and Gates. Carnegie viewed public schools and public libraries as being crucial for making knowledge and information freely available to individuals. While Carnegie idealized hard work, self-improvement, and self-reliance despite potentially punishing economic and material conditions, it was the publicly and freely supported immaterial labor (self-education) that the individual could pursue for self-improvement and economic advancement. For Carnegie, while the public sector should certainly not redistribute access to public control over capital, the public sector should make freely available the means for individual access to information that would benefit the individual and contribute to the making of a more educated workforce and informed citizenry. On the contrary, Bill Gates earned his historically unmatched fortune specifically by using intellectual property laws to own, control, and license the products of immaterial labor, namely, software and digital information. That is, Gates's wealth is principally the result not of the sharing and free exchange of knowledge in the public domain, celebrated as the route to freedom and a democratic public by Carnegie, but rather Gate's wealth is a product of the restriction and commodification of knowledge. In the 1970s, computer hobbyists freely shared their hardware and software innovations in a kind of hippy-tech movement. Some of the software that would go on to result in spectacular profits for Microsoft, Apple, and other computer companies began as freely shared innovations by hobbyists. Gates and Steve Jobs, among other early leaders of the nascent computer industry, were particularly adept at commercializing and monopolizing the innovations of others.[11] In fact, what would come to be called shareware or open source is closer to the spirit of the early software and computer innovators who were motivated less by the potential for profits than they were by intellectual curiosity, the technology itself, and the challenges of solving problems.

While Carnegie eschewed conspicuous displays of wealth and excessive consumption, Gates champions a version of schooling that idealizes a corporate economy in which consumer spending on manufactured needs is at the core. So, VP intensifies the economic rationalization of giving by insisting that giving be more tightly controlled, especially in terms of its outcomes. Yet, it also departs from the ties that scientific philanthropy had to the ideals of a productive industrial economy. In a sense, the transformation of philanthropy reflects the transformation in the understanding of productivity and utility accompanying the shift in the United States from an industrial to a consumer and service-based economy. To put it differently, as the core of the economy has become increasingly defined by the imperative for economic growth dependent on ever more frivolous consumer spending and the fabrication of ever new irrational consumer needs and desires, unplanned and unrationalized giving appears increasingly as a problem in need of eradication. One way to think about this is that as squandering and irrational expenditure of energy, wealth, and resources is increasingly central to economic growth in a consumer society, squandering and irrational expenditure, like the giving of charity, appear increasingly as a problem, and it must be rationalized and expressed through authorized, legitimate, planned, and orderly forms. Charity must appear as investment. The logic of this creeping rationalization in the irrational consumer economy is particularly evident in the aesthetic realm. A value on and celebration of utility, the display of usefulness can be seen, for example, in the long-standing popularity (from the early 1990s to the present) of the SUV in which the individual sporting (display) of potential usefulness is both a kind of status and carries with it a moralism about excessive expenditure—the luxury car is decadent, but it is acceptable when in the form of a useful truck. Of course, in reality what could be a more frivolous expenditure than using a large gas-guzzling truck to do errands around town? This became obvious only when the price of gas radically spiked to over $4 per gallon in a short span of time in 2008.

The origins of the shift from scientific to VP in education thoroughly coincides with the broader neoliberal shift in education that can be traced back at least to the landmark report of 1983: *A Nation at Risk*. The report, *A Nation at Risk*, was significant for reframing the animating ideals of public education through national economic competition and declaring a "rising tide of mediocrity," prompting

hundreds of task forces to reevaluate every aspect of public education in the terms of the role education could play for global economic competition and workforce preparation.[12] Richard Lee
Colvin writes,

> Demographic, technological, and economic changes in the United
> States and worldwide had made education more valuable for indi-
> viduals, and for the nation's economic well being, than ever before.
> It was that last development, brought into focus with stories of out-
> sourcing jobs and the multiplying number of engineers and scien-
> tists emerging from universities across India and Asia, that motivated
> Gates and others to remain committed to addressing the problems
> of the public schools.[13]

Conservative writers such as Colvin, Hess and others recognize the
shift to a service economy behind the redefined purpose of public
education for the economy. However, they affirm such an understanding of public education primarily in the service of a corporate
dominated service-based economy and accept the neoliberal assumptions that schooling should primarily serve an economic function.
They do not consider how such an understanding redefines the
public dimensions of public schooling in privatized ways.

The shift from scientific to VP in education has material and
ideological origins. The material origins include not only the concern by corporate leaders with schools, preparing workers for their
industries. Peter Frumkin recounts one narrative of the origins of
VP. This narrative explains VP as in part a function of the sudden
wealth created by the technology boom of the early 1990s before
the dot-com bubble burst. So-called Microsoft millionaires were
created through the corporate policy of granting stock options.
According to this narrative, these newly monied do-gooders were
entrepreneurial and bent on doing philanthropy in ways familiar
within their moment in business. This meant that venture capital,
utterly central to the dot-com boom, would become the model for
the new philanthropy. Hence, the importation into philanthropy of
"leveraging investments," retaining controls over the venture
capital.

Another narrative explaining the rise of VP in education focuses
on the case of the Annenberg Challenge. Walter Annenberg, a
media magnate, devoted his last years to giving away a large portion
of wealth, in particular, to other educational causes. He endowed
the Annenberg Schools of Communication and established the

Annenberg Challenge that gave the largest ever philanthropic donation to public schools in 1993 of $500 million. Districts were
required to provide matching funds, and these amounted to another
$600 million.[14] Comically, Frederick Hess, Richard Lee Colvin,
and others describe Walter Annenberg as an ambassador and philanthropist. While it is true that Annenberg was ambassador to the
United Kingdom under Nixon, Hess and company fail to mention
that the bulk of his fortune was made from selling copies of his junk
culture creations *TV Guide* and *Seventeen Magazine*—that is, effectively making a fortune by educating the public in values of vapid
consumerism.

According to conservative critics, the Annenberg Challenge was
a turning point in philanthropy discourse. According to Colvin,
venture philanthropists and others are intent on avoiding the outcomes of the Annenberg Challenge. They contend that a large
amount of money spent on reform resulted in gains, but that these
were inadequate because their goal is to radically transform educational governance.[15] According to Hess, "leveraged strategic giving" is a reaction to Annenberg and explains the venture approaches
of Gates, Broad, Walton, and Milken.[16] Venture philanthropy in
this view is an attempt to exert greater control by givers over the
money spent on reform and the process of reform itself. Colvin
admits that the Annenberg Challenge resulted in much needed
funding for teacher professional development, music, dance, and art
lessons, among other improvements and yet concludes that because
the public schools in question still "remain beset with difficulties";
only the radical redistribution of control over public schooling can
effect necessary change. This argument is symptomatic of how, in
this literature, investing in public schools even when successful is
never adequate and so privatization can be the only route. What
goes unexplained is if these funding increases have been demon-
strably successful, why they should not be amplified and expanded
or made the basis for arguments for greater universal funding.

In fact, the Annenberg Report not only announces the successful implementation of increased funding, but it calls for adequate,
equitable, and reliable funding. Venture philanthropy will achieve
only the opposite by pushing public schooling toward functioning
like a market. It will necessarily bring the same inequalities of access
and choice that structure markets. For example, the expansion of
voucher schemes will only worsen the ways that inequalities in
wealth and income determine educational opportunities as

"consumers" of educational services with greater capital will add to the public vouchers. In fact, this is precisely what is happening with the ways that Local Educational Foundations (LEFs) are working. These local philanthropies started by rich parents are subsidizing their own children's public schools, in effect furthering privatizing what is already structured through unequal funding by property wealth. These LEFs undermine universal public education as the quality and support of a public service gets more closely tied to the wealth of the recipient of the service.

If the Annenberg Challenge offers a narrative explaining and justifyingthe shift to VP, it is on the basis of institutional response grounded in values of philanthropic efficacy. Such arguments efface the politics of education, presuming that everyone can agree as to what educational quality means. These perspectives offer little by way of explaining the broader economic and cultural shifts giving rise to VP. This effacement of the politics of education in VP was on full national display as the Annenberg Challenge linked Barack Obama's run for president to his relationship with education professor William Ayers at the University of Illinois, Chicago. The McCain campaign sought to show that Obama had a close relationship with Ayers largely by virtue of the two having served on the Annenberg committee at UIC. This was intended to paint Obama as a radical with views akin to those of the former Weather Underground member. Of course, aside from Obama's politics being far from radical, the media spectacle of the Ayers link prohibited a serious discussion of Ayer's educational views or Obama's educational views, as well as the extent to which the Annenberg Challenge was attacked by the proponents of VP.

Hess, for example, criticizes the Annenberg Challenge for not being radical enough. This is quite amazing in the context of the presidential election rhetoric in which the Annenberg Challenge was framed as the smoking gun of Obama's alleged secret radicalism. In fact, Obama's education views do not appear similar to Ayer's emphasis on teaching for social justice and appear rather to be closer to the views of the Commercial Club of Chicago, its Renaissance 2010 plan, and CPS CEO Arne Duncan, and he significantly has expanded federal funding for charter school promotion. In fact, Randi Weingartner the somewhat conciliatory former head of the AFT has gone so far as to describe Obama's education plans as the continuation of the Bush education plan. As critics contend its overemphasis on test-based forms of accountability and its continuation of the privatization agenda looks like a direct continuation of the Republican education agenda. As the federal government under Obama follows the lead of the Gates foundation's charter expansion strategy, in 2008 the Gates Foundation announced it was moving on to new education territory. This ought to be recognized as precisely the kind of work that VP does in "leveraging" the public into acting in privatized ways. Nonetheless, the rise of VP in education is consumed with a focus on strategy, efficacy, and method, eschewing underlying concerns with the overarching values and goals of public schooling as well as with the contested ideological positions animating particular reform initiatives. Perhaps, nothing succinctly illustrates this better than the incessant call for "what works" from the Gates Foundation to the Broad Foundation to Obama himself. Works to do what? For what? Why?

Peter Frumkin, a leading authority on philanthropy, locates the origins of VP with an investment metaphor that entered political discourse with Clinton-era "new Democrats" in 1991.[17] The "new Democrats" headed by the Democratic Leadership Council (DLC) sought to steal issues from the political right: they embraced both neoliberal policies and a "post-political" "post-ideological" perspective. The neoliberalism took form with such reforms under the Clinton administration as follows: the attack on the liberal welfare state; the dismantling of welfare and creation of punishing workfare; the support for privatization; and the contracting out of such public services as public housing, government functions, and public schooling; media consolidation and economic deregulation in terms of removing some of the limits on corporate growth and allowing commercial and speculative financialization to merge. The "post-ideological" perspective of the "right wing of the democratic party" was shared with the "third way post politics" of New Labour in the United Kingdom. The moment of the rise of "post-ideological" politics came on the heels of the collapse of the Soviet Union and with it the departure of a sense of some outside or alternative to liberal democratic electoral government paired with *lazier-faire* capitalism. Margaret Thatcher's "TINA thesis" "There Is No Alternative") to the market became the view of the new democrats as a triumphalist rhetoric took hold throughout the nineties, heralding "the end of history" and with it a beneficent corporate-led globalization. Trade liberalization and the privatization of public services were pushed for all domains as increasingly the state became framed as nothing but bureaucratically encumbering and in need of radical market-oriented reforms. In education, this also marked the entry of the neoliberal education Bible in the form of John Chubb and Terry Moe's *Politics, Markets, and America's Schools* that elaborated the neoliberal gospel applied to public schooling. The market, it was argued, is more democratic than public deliberation because the market is inherently more efficient than the public. Markets and democracy were not distinguishable. Part of what this view missed is that markets function as political systems, producing social and class hierarchies, and businesses do not operate on democratic principles but rather on authoritarian principles. As Frumkin illustrates, part of the "New Democrat" turn to neoliberalism involved a redescription of social values and commitments in market terms. "New Democrats" jetissoned the language of "higher taxes" and "spending" and employed the language of "contributions" and "social investments."[18]

Frumpkin describes as a "marriage made in heaven" the linkage of Silicon Valley venture capitalism and the rhetoric of the "New Democrats." Central to this metaphorizing of public provision as business was the implicit assumption of a naturally disciplined and disciplining market. Frumkin writes,

> Rather than simply being a purvey of charitable funds for deserving organizations of all sorts, venture philanthropy promised to turn donors into hard-nosed social investors by bringing the discipline of the investment world to a field that had for over a century relied on good faith and trust.[19]

According to Frumkin, VP's origins have much to do with the interest of new technological entrepreneurs in developing measurably effective outcomes for their "social investments" and to "transfer wisdom across sectors." Such a transfer of wisdom, however, carries with it certain social costs as a publicly oriented philanthropy becomes increasingly understood through the private lens of profit accumulation. This is no more evident than in the ways that VP transforms educational governance and reinvents the liberal democratic value of pluralism.

There is wide agreement across the political spectrum that the leading VPs, especially Gates and Walton, have extremely concentrated control over these educational privatization agendas and the numerous influential synergistic organizations that they fund.

> It seems safe to estimate that the (Gates) foundation is providing well over two-thirds of all philanthropic giving to high school

reform, creating the possibility that one donor will be largely responsible for the shape of modern day high schools.[20]

The implications of this concentrated power over policy formation have even the staunchest advocates of privatization worrying about the governance implications of so few people setting educational agendas that potentially influence so many. As proponents Hassel and Way write,

> While this concentration has enabled a strategic focus that would not be possible with a more far-flung group of funders, it raises tough questions about how healthy it is for a relatively small number of donors to shape the direction of school choice institutions so directly.[21]

Even Frederick Hess of the American Enterprise Institute, who aggressively defends the concentrated influence of Walton and Gates, writes "to observe the reality of this oligopoly is not necessarily to pass judgment on it." He nonetheless worries that the oligopoly could "stifle unpopular critiques, eliminate unconventional ideas, and shut out promising entrepreneurs."[22]

Venture philanthropists advocating public school privatization link the traditional philanthropy ideal of pluralism to the new emphasis on "leveraging change." Traditional philanthropy has been defended from regulation on the grounds of pluralism. That is, according to the claims for philanthropy's pluralist tendencies, a largely unfettered, unregulated field of philanthropic giving results in a broad array of philanthropic projects across the ideological spectrum, and this promotes a healthy diversity of social projects.[23] The liberal value on the "marketplace of ideas" suggests that givers "vote with their dollars" about what they want to support. As Frumkin contends, small donors account for about one-half of all philanthropic giving. For Frumkin and others supporting the pluralism defense of philanthropy, the diverse values of the people are expressed by their giving and the diverse needs of the people are met by a diverse array of gifts. So, for example, on the one side the popular will to defend or make abortion illegal would be represented by givers committed to these causes and, on the other side, recipients of their largess willing to open abortion clinics or form abortion protest groups and mount legal challenges, and so on. Likewise, from the perspective of the pluralist defense of philanthropy, on the one side, givers to the cause of educational privatization will make gifts to think tanks and grassroots organizations and political action committees that support such perspectives, and then, on the other side, there will be givers to anti-privatization forces such as grassroots organizations, progressive foundations, and so on. For Frumkin, philanthropy only facilitates the motives of different actors on a politically level playing field.

These examples highlight how limited an understanding the pluralist defense of philanthropy holds. First, philanthropic giving is frequently connected to broader political movements and ideological configurations. Public school privatization is part of a broader effort by conservatives to roll back or privatize public provisions and social spending. Public spending is fought over by different groups and different classes. Fiscal elites fight to redistribute public priorities and spending in ways that benefit those at the top of the economy. They have more material resources to wage such a battle. This becomes readily apparent with the vast amount of foundation wealth spent on the privatization agenda. These vast sums of money are not pouring in to create foundations to defend and strengthen public schooling or to lobby for the equalization of educational funding across the class spectrum. Rather, the educational lobbying that is done more often than not assures that educational resources are retained for those in positions of class privilege.

Another problem with the pluralist defense of philanthropy is that it misses how the ideological frameworks that inform how individuals understand issues such as school privatization are tied to material stakes. For Bill Gates, John Walton, Donald Fischer, Eli Broad, Michael Milken and other venture philanthropists, business appears as universally beneficial. After all, it has worked well for them. So business language, values, and ways of seeing should be applied to the public sector. Of course, to use Walton and Wal-mart as an example, business competition and the relentless drive to cut labor costs and avoid unions has resulted in cheap consumer goods in Wal-Mart stores but also low paying jobs, vast outsourcing, reliance on authoritarian governments in foreign countries to assure cheap labor, the denial of adequate benefits, and so on. These profit accumulation strategies have resulted in unparalleled wealth for the owners but new lows for workers' well-being, the decimation of small town business, and a race to the bottom for quality of work in China and elsewhere. Business assumptions are not universally beneficial. They are selectively beneficial for those at the top of the business. But public institutions have a mandate of being universally beneficial.

Perhaps nothing better illustrates the limitations of the argument for pluralism than Frumkin's discussion in his book *Strategic Giving* of VP in education. Here, Frumkin describes *Education Next* magazine, which is a product of the neoliberal Hoover Institution, as a nonpartisan effort merely dedicated to getting new ideas into the public forum. That Frumkin describes one of the most partisan pro-privatization publications as politically neutral suggests more than Frumkin's truncated perspective; it also represents a dire failure of the ideal of pluralism to comprehend the political nature of educational reform and the role of philanthropy in expressing competing political agendas.

Venture philanthropy is a bankrupt ideal in light of the collapse of the neoliberal market fundamentalism and its emphasis on markets as self-regulating and the expansion of the market model to all aspects of public life. The projects of applying business rationales to all aspects of public schooling ought to be recognized as a hangover from a speculative bubble economy, from neoliberal economic dictates and ideology. These business ideals and metaphors as applied to public schooling need to be not only dropped but replaced with a recovered public sensibility, a universal value for the public schooling as a crucial part of a democratic public.

## 4

# THE GIFT OF CORPORATIZING
# EDUCATIONAL LEADERSHIP: THE
# BROAD FOUNDATION AND THE
# VENTURE PHILANTHROPY VIEW
# OF LEADERSHIP

Venture philanthropies have sought to transform K-12 schooling
in part by radically transforming administrator preparation. No
institution is more aggressive in pursuing such changes than the Eli
and Edythe Broad Education Foundation. Its initiatives include
creating educational leadership training projects specifically
designed to recruit noneducator corporate, military, and nonprofit
leaders. The Broad Foundation also seeks to deregulate teacher and
administrator preparation programs that will take such programs
away from the purview of universities and allow for their privatization. They create scholarships to fund schools and students that
reward "achievement gains," emphasize standardized test-based
performance achievement tracking, and also create test databases
for long-term tracking of student test scores to direct educational
policy and determine the effectiveness of teacher and administrator
preparation programs. The Broad Foundation also works to expand
charter schools and "franchise" charter management organizations.
While on the surface these initiatives may not seem closely related,
they share a common set of ideals and a cohesive vision for public
schooling that can best be understood as an expression of the ideology of corporate culture/neoliberal ideology applied to
education.

The Broad Foundation is one of three largest venture philanthropies along with the Gates and Walton foundations. However,

of the three, Broad has by far done the most to transform the running of public schools by seeking to influence administrator preparation, the meaning and value of teacher and administrator quality, and school boards. The Broad Foundation was created by and is run by Eli Broad, who became a billionaire in the real estate and finance industries. Since his retirement, he has put his energies full-time into public school reform.

Broad's central assumptions about improving public schooling include: (1) that the problems facing public schools are administrative problems caused by bad management practices—especially caused by bad public school managers who lack the leadership skills of the private sector; (2) that public school improvement begins with top-down reform; (3) that educational quality can be understood principally through standardized test-derived achievement scores and that poor and minority students suffer from an "achievement gap," which can be remedied through better educational methods and management. Of course, on their own, many of these assumptions are widely held rather than specific to Broad. However, taken together, these assumptions are closely aligned with the neoliberal educational reform movement as championed by the Fordham Foundation, American Enterprise Institute, Hoover Institution and leading right-wing policy wonks associated with them, especially Chester Finn and Frederick Hess among others.

In what follows here, I focus on three aspects of Broad's educational projects to illustrate how what is represented in academic and public discourse as generosity, care, excellence, and improvement ought rather to be understood as an expression of particular values, visions, and political ideologies in education that are hostile to public forms of schooling, that celebrate and promote a corporate and private rather than public perspective on educational governance, and that have an anti-intellectual and anti-critical approach to knowledge and curriculum.

## THE LEADERSHIP AGENDA

A central priority of the Eli Broad Foundation is to recruit and train superintendents and principals from outside of the ranks of professional teachers and educational administrators and, related to this, to shift administrator preparation away from universities and state certification to the control of outside organizations that embrace corporate and military styles of management and that share the VP agenda. These programs include most notably Broad, New Leaders for New Schools, and the training program of "Knowledge Is Power Program" (KIPP). At the core of these initiatives has been the neoliberal celebration of private sector and denigration of all things public. In this view, educational leadership is imagined ideally as corporate management, and the legacy of public educational administration is devalued. Policy literature in the area of educational administration refers to what Broad spearheads for leadership as the "deregulation agenda."

BetsAnn Smith, who writes that the Broad Foundation is part of this movement to end certification and licensure in universities and create outside deregulated educational institutions for leader preparation, has done one of the most comprehensive examinations of the deregulation agenda. Smith contends that not only is there no evidence for the success of this deregulation movement that is being pushed by right-wing think tanks and corporate foundations, but that the turn to outside leaders relies heavily on what she calls a "compositional argument"—that is, a cultural narrative about the "bullish CEO." To put it more expansively, the call for turning to leaders from the business sector and the military should be understood not merely as one cultural narrative but as a cultural narrative that is part of a broader ideology of corporate culture within which a series of interlocking business and military metaphors plays a central role in setting the stage for policy.

In the case of the outside leader ideal, the educational administrator as "bullish CEO" merges with the description of educational values through metaphors of efficiency, choice, competition, and accountability. These metaphors rely for their intelligibility on their opposites including ascription of the public bureaucracy and the ensconced public leader as inefficient, monolithic and imposing, monopolistic, and unaccountable. The educational leader as "bullish CEO" hence participates in the much broader tendency, found across scholarly and public discourse, to imagine the school as a business, the school workers as business people, the student as consumer of private services. Within this view of privatized schooling, the leader should naturally be from the private sector or from the military.

Within this corporatized view of educational leadership found in VP, military leadership is celebrated for its alleged link with corporate management—a focus on discipline, order, and enforcement of mandates through a hierarchy at every level of public schooling.

The "natural discipline" of the market is discursively linked to the corporeal discipline of the military. The turn to military leaders of public schools began in the late 1990s with Seattle and Washington, DC, appointing military generals as "CEOs." This has picked up speed, as seen in the expansion of programs such as "Troops to Teachers" that puts veterans in the classroom, the expansion of public schools run as military academies (Chicago leads the nation with six schools so far), increases in military recruitment in schools accompanied by slick corporate youth advertising, and the NCLB law that mandated student personal information would automatically be given to military recruiters unless parents intervened. The turn to military leaders particularly for the urban poor and predominantly African American and Latino student bodies belies a profoundly racialized phenomenon within which these students are framed as suffering primarily from a lack of discipline, which the military and the corporation can supply.

The discourse of discipline typified by the turn to the military and corporate leader actively denies the social conditions informing the experience of schooling. Instead of acknowledging how social inequalities influence educational access, such discourse reduces the language of educational opportunity to a narrative of individual discipline. Broken schools, absent textbooks, underpaid and overworked teachers, large class sizes, communities beset by unemployment, public disinvestment, dire poverty, skyrocketing homelessness, not to mention unequal distribution of cultural capital—in short, all of the material and symbolic social conditions inside and outside of schools that render schooling difficult to impossible are made to seem as irrelevant when discipline frames schooling. The celebration of the disciplinarian administrator is deployed in conjunction with multiple other disciplinarian policies such as the implementation of school uniforms, zero tolerance policies for expelling students, vast expansion of surveillance technologies in schools, surprise searches, and police school invasions.[1] The turn to the authoritarian disciplinarian can be found not only in policy but across public discourse in such popular films as *Stand and Deliver, The Substitute, 187,* and it participates in what Henry Giroux, Mike Males, Lawrence Grossberg, and others have extensively detailed as a discursive and material "war on youth" waged in the United States. In this "war," kids are blamed for a myriad of social and economic problems while legal and public protections for kids are scaled back.[2]

Through most of the first decade of the new millennium, an unabated barrage of representations across mass media-educated Americans in the virtues of the hard-nosed CEO, from Jack Welch and his goal of regularly firing ten percent of the General Electric workforce to discipline the entire company to the return of an omnipresent Donald Trump selling viewers the fantasy of being an apprentice bullish CEO on reality TV. In this context, the billionaire CEO Eli Broad and his application of business ideals to educational leadership appeared as offering the gift of corporate and military efficiencies and also discipline to the beleaguered public schools. But, the context for interpretation has recently changed.

As the financial crisis of 2008 hit, it became readily apparent across the political spectrum that the neoliberal idea of markets regulating themselves without state support and intervention is no longer tenable. (In a sense, it never was very credible, as the neoliberal program required state support and regulation despite the ideology.[3]) As waves of financial corporations collapsed or had to be saved through massive federal intervention, the assumptions behind the market fundamentalism of the last 40 years began to be called into question. Neoliberal former head of the U.S. Federal Reserve bank Alan Greenspan appeared before Congress and admitted that his decision, based on his view of the economy, to allow greater and greater deregulation of derivative markets had been wrong because, in his words, "I found a flaw in the model that I perceived as the critical functioning structure that defines how the world works."[4] Liberal economist Joseph Stiglitz came to a similar conclusion that the financial crisis could best be understood by grasping that everything came back to deregulation and the faith in markets to regulate themselves. As Henry Giroux and Susan Giroux wisely point out, a consequence of the broader pedagogical effect of neoliberalism on both education and the culture at large has been a difficulty for the public to formulate and name alternative visions to the failed neoliberal ones. This is, as they rightly suggest, a significant problem with a legacy of schooling overtaken by anti-critical approaches such as standardized testing, scripted lessons, commercialism, pay for grades, and so on because they prohibit the kinds of questioning, critical dialogue, tools of investigation necessary for the fostering of democratic culture that citizens must learn in order to participate in reworking civil society with others.

The Broad Foundation's neoliberal approach to educational reform must be viewed with profound skepticism for two primary reasons. First, it is modeled on the same neoliberal assumptions (privatization and market deregulation) that have been thoroughly discredited as behind the economic crisis. In other words, Broad and the other venture philanthropists assume that education is like business and should adopt the same framing language and guiding rationales. Also, neoliberal ideology forms the foundation for expanding the market metaphor to all areas of social life, conflating public and private spheres, and eradicating a sense of the public good in favor of a society composed of nothing but private consumers.

Second, in the case of Eli Broad, at issue is not simply that he, like the other venture philanthropists, adopted the language of venture capital and sought to apply it to education. Broad's fortune and hence his ability to steer educational reform, debate, and policy through his foundation all derives from the two primary industries at the center of the financial crisis and subsequent economic meltdown—namely, real estate and finance. What is more, Broad, made a killing in these industries specifically by using them synergistically rather than competitively. Therefore, the Broad narrative of financial success, deregulation, and idealization of corporate culture falls apart not only due to the collapse of the neoliberal ideology that grounds it but also due to the fact that Broad's neoliberal educational reform was always premised on assumptions that contradicted the origins of Broad's own wealth—namely, speculative capital in a bubble economy and monopolistic behavior. What the financial crisis reveals about Broad is that what he has sold as a narrative of skill and hard work that every school child should emulate ought to be understood as the result of the clever working of an economic context that was grounded in layers of gambling—an economy that was a speculative house of cards.

The mortgage crisis of 2008 was the result of deregulation of banking compounded by the hawking of mortgage-backed securities that when made into securities were sliced and diced to sell what appeared as secure assets but were in fact highly risky and speculative. This in turn was compounded by the linkage to these mortgage-backed securities of loan default swaps that were effectively free insurance policies on the mortgage-backed securities that multiplied the effect of debt creation when these bad investments failed. The amount of money in the economy multiplied radically through speculation between 2001 and 2007. At their most basic, Broad's fortune was based in speculative capital made possible by the neoliberal dictate of deregulation. The removal of public controls over private capital set the stage for the amassing of Broad's personal fortune, and it participated in the broader radical upward redistribution of wealth and income throughout the last 30 years.

It is precisely the neoliberal ideal of deregulation, a form of class warfare waged by the rich against the rest that Broad extrapolates as a metaphor to apply to public education. If only the public sector can be made to look like and act like the private sector, so goes the metaphor, then the public sector can be improved. And the only way to do that is to shrink public control over public institutions and hand control over to those from the private sector. But, the metaphor is misleading. There are great differences between public and private institutions, their missions, and their leadership.

Educational leaders for public schools have distinctly public obligations and responsibilities that differ from the obligations and responsibilities of private sector managers. Private sector managers are responsible foremost for maximizing profit for owners or shareholders. Their decision-making, skills, training, and relations with employees, in short, all that a private sector manager does, foster the financial goal of profit. Educational leaders for public schools are responsible to the public, namely to the community, the parents, students, and teachers who form it. The end of public school administration is not profit maximization but public service. Additionally, the private corporation has a particularly hierarchical organizational structure with the owners and managers at the top with near absolute authority. The public school, being publicly accountable, has a considerably more democratic organizational structure with administrators answerable to multiple constituencies within the community.

In both a practical and political sense, as BetsAnn Smith suggests, the outside private manager will not be attracted to the kinds of programs championed by Broad because though "[a]ttracted to the idea of 'running a school,' many aspirants overlook public schooling's democratic complexities and the degree to which its leadership demands are unrelenting and unrelentingly public."[5] Smith's study of the case of deregulation in Michigan highlights the differences between large districts that have been subject to anti-democratic mayoral takeovers as opposed to the majority of districts that remain subject to democratic oversight. The mayorally appointed CEOs sit in closed-door meetings where they are generally told that the priorities are test scores and "restoring fiscal order."[6] Such setups subvert the messy public struggles for educational and public priorities waged in and through public institutions. Of course, in the context of an era of high-stakes standardized testing and the standardization of curricula and other anti-critical approaches to teaching and learning, such narrow imperatives for test scores and cost-cutting promoted in the name of business efficiency above all else become instruments to assure a profoundly anti-intellectual pedagogical approach to schooling dominated by rote learning and memorization, scripted lessons, and decontextualized fact. Not only do these approaches undermine the possibilities of public schools operating as critical intellectual public spheres, but they also have implications for the kinds of social relations, identification, and identities that they produce for the activities people do outside of schools. In other words, democratic culture depends upon the built capacities for criticism, debate, and deliberation that critically intellectual public schools can develop. The corporate approach to schooling of which the corporate bullish CEO is a part undermines the civic possibilities that public schooling can have for communities by imagining school governance as being imposed from above and outside rather than from within the community while suggesting that knowledge must be imposed and enforced rather than beginning with the experience of those in the community. Some educational policy writers on the political right are quite explicit in championing the corporate-style outside leader.

In the journal *Educational Policy*, Frederick Hess and Andrew Kelly of the neoliberal American Enterprise Institute call for a "radical" restructuring of leader preparation that would involve thoroughly importing business management and principles into the curriculum, redefining the meaning of candidate qualifications to be understood through the outside leader and primarily the business leader, stripping control of universities in leader preparation and also licensure and teacher education generally, and shifting control to foundations with VP ideals such as Broad, New Leaders for New Schools, and KIPP.[7] Hess and Kelly see as progress, though insufficiently "radical" progress, such programs as those state-based ones in Ohio and Georgia that are modeled on corporate management academies. The Ohio Principals Leadership Academy was run by a former trainer who developed management academies for companies such as Bath and Body Works.[8] But, for these authors, the problem is that they do not recruit enough middle management directly from business.

This outside leader ideal, as it is championed by Hess and Kelly, Finn, and others of the neoliberal perspective, calls for educational leadership candidates to be educated the way that New Leaders for New Schools does it: having a "proven" track record of leadership experience outside education before even beginning, candidates will then be further educated by business school and education faculty and will learn educational research and "business school literature on organizational change, management, negotiation, and conflict resolution."[9] Hess and Kelly also suggest that KIPP's corporate management training model for leadership preparation is ideal. It is housed in UC Berkeley's Haas School of Business and funded by the Fisher Foundation which is an aggressively pro-privatization VP foundation run by the owners of the Gap, Banana Republic, and Old Navy. Students learn from business professors while "the curriculum fuses the KIPP ethos that results matter with more conventional business practices."[10] "Through the examination of case studies about successful companies, such as Southwest Airlines and FedEx, students consider what lessons the private sector may hold for K-12 management."[11]

Like the corporate approach of New leaders for New Schools (NLNS) and KIPP, Broad's Academy is based in the ideal of making "great leaders." And great leaders for Broad come largely from business or accept a business view of administration. "Dan Katzir, the Broad Foundation's managing director and an instructor in the academy, told the fellows that Southwest Airlines and the computer giant Dell Inc. are examples of how new players entered an established market, came up with innovative strategies and achieved success. The message: Urban superintendents can, and should, do the same."[12] The point not to be missed here is that "great managers" for Broad follow the management style, precepts, assumptions, and language from the private sector. Part of what is at issue here is the VP approach to educational leadership that views public schools as a private market and that views private corporations as the model for public institutions. The confusion between public and private institutions and values has enormous implications for educational governance, material struggles over educational resources, and especially the conceptualization of knowledge in both public schools and in educational leadership programs.

This is not merely a matter of instituting a corporate style of educational leadership. Also, such pedagogy involves teaching future leaders to understand their identities in reference to the private sector rather than to the public sphere and teaching future leaders about the alleged virtues of privatization schemes such as "choice" and charter schools.[13] For example, such projects encourage social relations forged through the hierarchical organizational form of the corporation and the concentrated authority of the corporate leader rather than through the collective, dialogic wielding of power found in more democratic organizational forms.

## THE BROAD PROJECTS SHARE A VIEW
### ABOUT KNOWLEDGE

For Broad, public schools, teacher education programs, and educational leadership programs are all described as businesses. The description hangs on a metaphor of efficient delivery of a standardized product (knowledge) all along the product-supply chain: the product is alleged to be high quality, neutral, universally valuable education. The deliverable, knowledge, is positioned like product. In the case of K-12, knowledge, which is presumed to be universal and objective, is to be standardized, measured, and tested. Test scores in this view are the ultimate arbiter of educational quality and, like units of commodity or money, can allow for the quantification of growth and progress. For Broad, this is called, "achievement." Hence, one of Broad's major initiatives is the "closing of the achievement gap" and the funding of school districts and schools that raise the test scores of nonwhite students. The presumption is that the unequal distribution of the product, knowledge, can be remedied through methodological interventions such as the introduction of rigid pedagogical reforms, the introduction of proper business incentives such as teacher bonus pay, or payment to students for grades, as well as management reforms such as installing business people to manage schools, and getting unions and school boards out of the way of these business-based "achievement oriented" reform efforts. The moment the goal of education becomes "achievement," the crucial ongoing conversation about the purposes and values of schooling stops as does the struggle over whose knowledge, values and ways of seeing should be taught and learned.

This perspective about knowledge as measurable, quantifiable, universally valuable, and neutral direct Broad's biggest initiatives: the leadership agenda, the "Broad Prize," and the database project. The leadership agenda imagines educational leaders as business managers who can increase test-based achievement like increasing financial revenue and can decrease the "achievement gap" like a CEO seeking to close the earnings gap with the business competition.

As Fenwick English argues, the movement in educational leadership to standardize a knowledge base (and then enforce it through ISLLC/ELCC standards applied in NCATE/National Policy Board for Educational Administration) destroys the most valuable dimension of intellectual preparation offered by university leadership programs and effectively lowers standards in educational administration preparation by encouraging the proliferation of weak programs that offer advanced degrees. English contends that Broad typifies the "back door" to "a neoliberal global policy agenda to privatize educational preparation as advanced by right-wing, corporate-backed think tanks and foundations that proffer that free market approaches (i.e., marketization) are a better way to prepare educational leaders."[14] What makes these programs weak for English is that they are based in standards that form a "knowledge base." The standards: represent a mistaken view of knowledge as static rather than dynamic. Such standards are anti-democratic, work against historically marginalized groups, and secure the authority of those with the most political power; they are grounded in a "knowledge base" that functions primarily to exercise political power; they are ahistorical and conceptually incoherent, representing "disembodied skills, concepts, and ideas distanced from the theories that spawned them" that artificially ground existing relations of power; they are antiscientific and anti-intellectual, denying the necessity for research beyond what is encoded in the standards. Together, these problematic underlying assumptions set the stage for a radical VP transformation of administrator preparation defined by market competition, with students shopping for the most convenient certification program. Meanwhile, vacuous programs, lacking in intellectual rigor, proliferate. As English argues, the NCATE standards make the challenging of the assumptions of the field irrelevant or a problem for professors of administration. These standards as currently conceived do not conceptualize the field as dynamic and contested.

The bad assumption of a standardized knowledge base results in the stunted intellectual development of the field of administrator preparation. Consequently, the university's role in intellectual or critical preparation appears tenuous as off-campus "on-site" preparation programs rapidly expand. The theoretical and intellectual content of administration preparation shifts largely to efficacy-oriented literature from business management, or it is evacuated altogether. As well, this plays into the long-standing confusion in the field of education over the relationship between theory and practice as on-site learning takes precedence and an anti-critical practicalism takes over. Practice is positioned as the real stuff of administrator preparation grounded by the ultimate goal of "changing outcomes" measured by "increasing student achievement," which means raising test scores. Of course, such a conceptualization of educational leadership through the static knowledge base conceals who makes the content of such tests and the symbolic and material interests tied to such claims to truth. The positivism of this approach to knowledge separates facts from the assumptions, values, and ideologies that inform the selection and arrangement of facts.

The Broad approach typifies a much larger movement across the field of education to tie the preparation of teachers and administrators to the test outcomes of candidates' students. In other words, the value of an education professor or a person preparing teachers can be boiled down to the test scores of the student's student. The Carnegie Corporation has recently championed such thinking, and one variation of it goes by the title "value-added" education. The measure of the value of preparation programs is the "value" they have in upping scores. This way of thinking about teacher and administrator preparation exemplifies this resurgent positivism and its anti-intellectual bent. In this view, there can be no place for educational study that does not result in test score improvement two levels down. So educational theory, sociology and philosophy of education, curriculum theorizing, pedagogical theory,—approaches in education that address the underlying assumptions, ethical, historical, and political aspects of what is taught and learned—none has a place in the "value-added" perspective because all that matters is "delivery" of "content knowledge" through the use of the "best" "instructional methods."[15] As English rightly contends, educational leadership instruction and knowledge ought to be dynamic rather than static and ought to link research into educational problems with research into social problems.

## THE SCHOLARSHIP AGENDA

Broad created the Broad Prize for Urban Education. The foundation claims that what it has promoted as a "Nobel Prize for education" is intended to support public schooling and increase confidence in public education. The media has picked up the mantle of "Nobel prize for education" from Broad, and this has been endlessly repeated in the popular press. Broad divides a million dollars among five urban school districts that it has deemed as making improvements in "student performance" and "closing the racial achievement gap." One winning district gets $500,000 and four runners-up get $125,000 to be used for university scholarships for graduating seniors. Broad evaluates urban districts for the prize money by looking at state standardized tests, graduation rates, SAT and ACT scores, among other national tests.

Broad's educational reform agenda applies the same assumptions to rewarding schools and students as it does to training principals and superintendents. While in the case of the administrator there is a "knowledge base" that can collect and apply knowledge regardless of social context, in the case of promoting a particular school policy through the prize, it rewards standardized and largely decontextualized knowledge that is alleged to be of universal value. In this perspective, those students and schools that do not score highly on the standardized achievement tests can be "incentivized" through the promise of scholarship funds. One of the most obvious basic problems is the fact that the scholarship prize does not address the skyrocketing costs of higher education and that higher education could be publicly funded. The prize assumes that all students somehow will be able to afford to go to university. There are numerous other problems with this line of thinking that animates the scholarship.

First, the prize assumes that genuine learning should be measured principally by standardized tests composed of knowledge formulated by specialists. Second, this assumes that the tested knowledge is of universal value and expresses no class, cultural values, or perspectives, and it should be of universal interest. When poor and nonwhite students score poorly on these tests, Broad frames this as a deficit in those poor and nonwhite students. Such a deficit ought to be remedied by figuring out how to raise test scores. Broad's perspective runs contrary to more critical approaches to teaching and learning. According to these perspectives, learning ought to begin with what the learner knows; that student experience which is meaningful should then be problematized in relation to broader questions and problems to help students develop a greater understanding of what has produced the student's experience. This means that the student in a more critical approach must learn to approach knowledge not as decontextualized bytes to be unthinkingly consumed and regurgitated but in ways that comprehend that knowledge and versions of truth are struggled over and that, different interpretations are informed by material and symbolic power struggles. In this critical perspective, learning as an act of interpretation must be understood as inevitably linked to acts of intervention in the sense that there can be no neutral interpretations and that how students come to see the world informs how they act on the world. Rather than primarily developing the tools for repeating official knowledge from a critical perspective, students must learn to analyze what they come to know from experience and texts in social, political, ethical, and historical ways. Crucial questions at the center of this critical approach include who is making this knowledge, why do they claim this, how are these claims related to the position of the claimant, what kinds of broader structural forces inform the claim to truth? Broad's rewarding of knowledge that is foremost confirmable through standardized tests denies all of these crucial questions and shuts down the critical approach to knowledge.

Nothing better illustrates the stakes in these different approaches to learning than Broad's own involvement in educational reform following Hurricane Katrina in New Orleans. When Katrina devastated the city and its schools, long-standing politically failed privatization plans were put in place (spearheaded by right-wing think tanks like Urban Institute and Heritage), including vouchers and charter schools, the carving up of the school district, the dismantling of the teacher's union, and the refusal to rebuild the public schools as part of the business-led *Bring New Orleans Back Commission* (BNOBC) plan to keep poor and predominantly African American citizens from returning to their communities, homes, and schools. New Orleans was an experiment in neoliberal urban rebuilding.[16] The fate of the schools was struggled over. The history of systematic disinvestment in the New Orleans schools, the history of white flight, the failure of the corporate sector to contribute adequately to the public schools—these histories were conveniently erased as the Broad Foundation and the other venture philanthropies including Gates, Fisher, and Walton showed up to offer their generosity. The cash on hand came with strings. Rather than supporting the rebuilding of the public system in a better form, the venture philanthropies targeted their money at the creation of charter schools, alternative administrator preparation of the sort already discussed, and Teach for America program that expands an uncertified, undereducated, and deunionized teacher workforce. These initiatives, totaling $17.5 million, contributed significantly to the carving out of an elite charter network in the city for more privileged residents, solidifying the dispossession of predominantly poor and African American former residents, and the continuing attack on the teachers union.[17]

The kinds of "achievement oriented" standardized schooling such gifts foster will prohibit the kinds of critical teaching and learning that would encourage students to understand how Katrina, the city, and the schools became disputed terrains of class and racial struggles. Indeed, if public schooling is to offer democratic possibilities, such critical knowledge becomes crucial for students to have the skills to engage as public citizens in the formation of both community and knowledge making about it. Eli Broad's own words help to illuminate this. Speaking of Miami-Dade, Florida Broad said, "Miami-Dade is doing what some say is impossible—improving students' performance, regardless of their race or family income—while at the same time closing persistent achievement gaps."[18] Broad frames class position and cultural difference as needing to be erased or seen as impositions to students learning the right knowledge. What this view completely misses is how the school rewards the knowledge and cultural capital of students of class and cultural privilege while disaffirming the knowledge of students of oppressed classes and cultural groups.

Third, while Broad claims that the prize increases confidence in public education,[19] it undermines many of the public aspects of public education. One of the ways it does so is by misrepresenting knowledge as already discussed. However, it also "de-publicizes" public schooling by suggesting that the private businessman should have the power to designate and influence the determination of what is valuable knowledge for students to learn. Furthermore, this private businessman uses a series of private for-profit companies such as MPR Associates, Inc., to manage the prize selection process and another company Schoolworks, an educational consulting firm, to do site visits and collect information on prize candidates. Rather than the values of a community guiding reform, the values of the billionaire and the private educational consultants do. The very idea that the value and vision of public education should be steered and influenced by one who can fund the "education Nobel prize" aligns the values of learning less with enriching individual lives and collective social purpose defined by the love of learning or the social implications of it than with learning for extrinsic rewards, possessive individualism, and even celebrity adoration. Should the point be missed, Secretary of Education Margaret Spellings announced a $125,000 Broad award in Bridgeport, Connecticut, " 'This is like the Oscars for public education,' she beamed." Of course, mass media sells products by offering celebrity identifications and educating viewers to imagine themselves in celebrity relationships. As Zygmunt Bauman discusses, the cultural pedagogies of new media, TV, and film beset us with the problem of subjectivity fetishism—a world of people subjectified as commodities who misunderstand their commoditized selves as authentic and free of the market. What public schooling as a public site can offer us in this context is one place where commercialized forms of address and modes of identification can be criticized and where noncommodified versions of selfhood and values can be taught and learned. Not only does the Broad prize contribute to an expansion into public schools of the commercialism found nearly everywhere else in the society, but it also promotes the kind of learning in formal schooling that does not foster interpretation and questioning of commercial pedagogies that promote a privatized and individualized society outside of schooling. In other words, forms of schooling that make central social, cultural, political problems in the world ought to be brought into formal schooling while the test-oriented pedagogical approaches that Broad supports do just the opposite.

Fourth, Broad's funding of scholarships for students to go to college obscures some crucial public policy issues regarding public funding for higher education, the skyrocketing costs of tuition, and the increasing corporatization of the university. Rather than advocating for a greater role of the federal government in funding universal higher education, Broad instead promotes an exclusionary and lottery-style system of funding that resembles social Darwinian reality television programs like Survivor. While the aim of providing some students with access to college appears to be an admirable one, what needs to be recognized is that the Broad Foundation's actions function in sanctioning and legitimatizing a highly exclusionary system of access to higher education. As Stanley Aronowitz wisely writes criticizing one of the crucial precepts of progressive educational philosophy (specifically Dewey),

under the sign of egalitarianism, the idea [is] that class deficits can
be overcome by equalizing access to school opportunities without
questioning what those opportunities have to do with genuine education...The structure of schooling already embodies the class system of society, and for this reason the access debate is mired in a web
of misplaced concreteness. To gain entrance into schools always
entails placement into that system. "Equality of opportunity" for
class mobility is the system's tacit recognition that inequality is
normative.[20]

What Aronowitz means by genuine learning is what, drawing on
Hannah Arendt, he calls transmitting a "love for the world" and
"love for our children." He develops this to mean that radical imagination must stem from radical criticism. Instead, as Aronowitz
laments all too commonly, schools teach "conformity to the social,
cultural and occupational hierarchy" rather than the democratic
values that are often the official but unfulfilled principles guiding
schools. Broad's scholarship prize represents the reduction of the
possibilities of schooling to work and through universal schooling
*access to social mobility*, which, as Aronowitz points out, is not egalitarian at all.[21]

Broad's scholarship prizes promises equality through the potential of individual upward mobility through graduation. By setting
such bait, Broad fails to acknowledge the structural economic limits of job markets. The existing economy cannot globally accommodate good employment for a fully educated population. Consequently,
the real fulfillment of educational and economic uplift can only
come through collective action to change the conditions and standards of work to provide full employment at fair pay, security, and
so on. Students can be educated for such collective struggle rather
than for merely a compliance to the current economic arrangement
of "casino capitalism."[22]

## THE DATABASE PROJECT

Broad supports the School Information Partnership and the more
expansive Data Partnership. These are efforts to compile, track,
aggregate, and analyze student test scores with the long-term goal
of influencing school and teacher education policy based in the
data. The Education Department provided $4.7 million, and $50.9
million came from private organizations. Of this, Broad provided
half of the amount. One explicit goal is to foster the aims of No

Child Left Behind (NCLB) to provide schools and parents with score information. The literature on the database projects suggests that data offers parents and students some information for "school choice"—that is, privatization. As with the Broad Prize, the foundation's justification in supporting this project is to narrow the racial "achievement gap." This is also a lucrative opportunity for information technology companies including Data Partnership collaborators CELT Corp. which designs information technology systems for schools, the school evaluation division of Standard & Poor's, and the influential Achieve Inc., a nonprofit organization that was founded by corporations and governors to promote "standards-based" education. In addition to Broad, the Gates Foundation heavily supports the Database Partnership.

Mass data collection of student test scores appeals to many who embrace an understanding of learning through numerically quantifiable "standards" imposed from above. Longitudinal tracking of test scores appeals to those who want to boil down successful teaching practice to "efficient delivery" of curriculum. In this perspective, instructional methodologies become the primary concern of teacher practice, and methodologies are disconnected from the matter of what is taught. The experts who know determine what students should learn. The teacher becomes a routinized technician proficiently executing what has been determined to be the most efficient instructional methods to raise test scores. In the tradition of Taylorism's scientific management, the classroom becomes "teacher proof."

The database project aims to track and identify which teachers and teacher approaches raise standardized test scores despite racial, ethnic, class, or linguistic differences. Then, once the instructional methods that most raise test scores can be identified, the teacher education approaches that those test-improving teachers were exposed to can be replicated. The database promises to highlight which schools' methods result in raised test scores by minorities and hence purports to provide information that will enable administrators to work to "close the achievement gap." Another promise is that the schools that score the highest on tests will be attractive "choices" for parents. Hence, the database project appears to work in conjunction with the way that NCLB set the stage for transforming public education into a national market by requiring local schools to allow enrollments by anyone who chooses to go to the school.

What is wrong with the database project is that it reduces the value of schooling to standardized test scores while effacing the ways that standardized tests correlate most closely with family income and cultural capital. The emphasis on instructional methodologies paired with the delivery of standardized units of curriculum rewards and promotes approaches to teaching that thoroughly ignores the social contexts within which students learn as well as the identities of the students. As a consequence, such approaches encourage teaching to be viewed not as an intellectual practice, nor as critical practice, but rather as a technical skill. The overemphasis on testing as the definition of student achievement has practical and social costs. It compromises pedagogical approaches oriented toward creative problem-solving while rendering pedagogical content estranged. Critical educators emphasize that learning ideally ought to begin with meaningful knowledge that students have experienced, that can be problematized in relation to broader social, political, and cultural contexts, and that forces to help students comprehend how their experiences, understandings, and assumptions are produced but also to help students theorize how to confront those forces that produce their experiences. The standards-based approach undermines critical confrontation with both student experience and with the social forces and actors that tell students what is valuable to know.

When it comes to policy, the database project lends itself to a positivist separation of facts from the values and assumptions that organize facts.

> Participating states will each receive a customized analysis of data needs and how to close any gaps from the CELT Corp. Over time, the partnership could help states build the architecture for a more robust data system, including detailed implementation plans, joint requests for proposals, procurement, and contractor oversight and management. CELT also plans to identify and share best practices across participating states.[23]

The point not to be missed here is that policy will be "data driven." In reality, data cannot "drive" policy. The very expression conceals the motives and politics under girding human beings' decisions about curriculum, pedagogy, teacher education, and administration. Implementing practices used in high-scoring schools in low-scoring schools will not only result in misapplications of pedagogical approaches, it also naturalizes the unthinking consumption of information as the essence of achievement while imagining teachers as little more than fleshy delivery machines. What does not get interrogated in all this is the process whereby some people with particular values, perspectives, and ideological convictions determine what is important for students to know. Belied is the question of who has the power to distribute and universalize this knowledge, whose material and symbolic interests it represents, who profits from it financially, and what is lost in terms of schooling as dialogic and intellectually dynamic. The database project promises inclusion and access. Yet, it is highly exclusionary by universalizing approaches to learning that refuse to engage with the different contexts that students bring to the learning situation. Context-based pedagogical approaches enable students who are traditionally excluded from the curriculum and who come from historically oppressed groups to problematize claims to truth in relation to their experiences. As in colonialist education policies, the learner must assimilate or perish.[24]

There is a grand irony in the data partnership. Its web site was developed by the National Center for Educational Accountability and Standard & Poor's school evaluation services division. Standard & Poor's along with the other credit rating agencies came under intense criticism in 2007 and 2008 as the collateralized debt obligations (CDOs) markets that S&P had rated highly collapsed. These CDOs are largely understood as triggering the broad-based global financial meltdown following their implication in the radical expansion of a speculative economic bubble.[25] As well, Standard & Poor's continued to rate the government of Iceland highly just up to its financial collapse in 2008. The point not to be missed here is that the database project in education is driven by a number of neoliberal assumptions that ought to be seen as thoroughly discredited. The unquestionable efficiency of business, the model of the numerically quantifiable progress derived in part from industry and the financial sector, the rating of students and teachers through quasi-credit ratings—all of these are called into question not by the financial crisis but by the failure of the neoliberal dictates that tell people to think of all social goods through the logic of economics. Rather than using dubious credit rating tactics to measure school children, teachers, and knowledge as if they were investments and commodities, human measures of the value of teaching, learning, and knowledge must be expanded. Perhaps as well, to turn it around, financial investments can be rated through their social values and social costs on a human index. Perhaps the United States could take a lesson in this regard from the leader of Bhutan who established as a leading social indicator, Gross Domestic Happiness.

# The Gift of Corporatizing
# Teacher Education and
# Higher Education

## Introduction

The earlier chapters have provided an overview and analysis of the major activities of some of the largest and most influential venture philanthropists and their various efforts to privatize and deregulate K-12 schooling and educational leadership. This chapter focuses on some of the ways that VP seeks to change teacher education and higher education generally. As in the other chapters, the central concerns here are with the questions of democratic process and governance, the inordinate power of venture philanthropies to influence, steer, and "leverage" public schools, the financial redistributive effects of VP steering, and the implementation of reforms derived from corporate culture in ways that undermine the democratic possibilities of public schooling.

In order to address the VP influence on teacher education and higher education, I look first to the current debates over professionalizing versus deregulating teacher education. I then turn to the broader contemporary debates over teacher education reform and highlight the competing ideologies at work in the current agendas. Specifically, I look to liberal, neoliberal, and critical calls for reform. I contend that the VP approach to teacher education and higher education, which squares most closely with the neoliberal view and is massively funded and promoted, undermines liberal, progressive, and radically democratic visions and values. I then recount the historical interventions of philanthropy in teacher education and higher education reform. It is important to foreground this history because both the professionalization and deregulation agendas refer to this history in their efforts. I then return to the largest VP, The Gates Foundation and examine its new focus announced at the end of 2008 on higher education preparedness. Of central concern in this chapter is the ways that VP initiatives collude with and foster the corporatization of teacher education and higher education just at the moment of economic and cultural crises that calls such an approach radically into question. What is at stake in big money pushing the neoliberal VP agenda is that it offers little in the way of coherently responding to the crisis of neoliberalism as ideology or economic doctrine, while undermining the public, democratic, and critical potential of teacher education. The Gates Foundation's new-found emphasis on influencing teachers and teaching as of 2008 symtomatizes VP's reduction of broad socially related educational problems to matters of individual teachers, and it wages a renewed assault on teachers, teacher unions, and teacher education. It is incumbent upon critical and progressive educators to reject the deregulation and professionalization agendas and instead develop and implement critical forms of teacher education that can be the basis for the broader expansion of democratic social relations and a different sense of social and individual obligation.

## THE POLITICAL STRUGGLE FOR TEACHER EDUCATION

In the first decade of the new millennium, teacher education has become increasingly subject to frenzied calls for transformation. As David Labaree has argued, teacher education has been historically subject to intervention from multiple constituencies in part due to its lowly institutional status, the deceptive appearance of teaching as easy when it is in fact difficult, and problems introduced by both consumerism and "administrative progressivism."[1] Most recently liberal scholars led by Linda Darling-Hammond have continued to push for the professionalization of teacher education in part by modeling proposed transformations on other professions such as medicine and law with the aim of raising standards and expanding equity.[2] Neoliberals have sought to transform teacher education by deregulating it from university and state control and privatizing it with the principle aim of what they see as better serving business and the national economy. While critical scholars have argued for a political understanding of the stakes in the teacher education reform proposals generated by liberals and conservatives, few criticalists have offered a program for changing teacher education. This chapter aims to contribute to such an effort.

Marilyn Cochran-Smith and Mary Kim Fries map the terrain of teacher education reform initiatives and suggest that the dominant debate over teacher education reforms can be characterized by, on one side, those advocating professionalization (liberals) and, on the other side, those advocating the deregulation of teacher education (neoliberals). The move for professionalization that is led by Stanford education professor and Obama education campaign advisor Darling-Hammond, calls for "a consistent approach to teacher education nationwide based on high standards for the initial preparation, licensing, and certification of teachers."[3] Efforts to expand the professionalization agenda are backed by traditional philanthropies such as the Carnegie Corporation, Pew Charitable Trust, Ford Foundation, and DeWitt Wallace Reader's Digest Fund.[4] The professionalization agenda involves implementing national standards for teacher education, instituting ongoing teacher professional development, and putting in place "performance based assessment" of teaching practice for the duration of the career. The clear emphasis in this approach is on improving teaching as practice and process while the emphasis on "performance based assessment" presumes measurable "achievement" and a linkage of test-based achievement to the reform of teacher practices.

The professionalization approach relies for its appeal on a connection between teacher preparation and professional career preparation such as medicine and law and the widely held beliefs as to the rigorous standards required of such programs. The comparison works in part by bringing the relatively highly remunerated private sector careers (doctor and lawyer) into comparison with the relatively poorly remunerated public sector career of the teacher. In fact, the advocates of the professionalization approach would for the most part like to see the status and pay of the teacher rise in accord with other professions.

These debates miss the extent to which the professionalization argument relies for its force on a largely unacknowledged reality. That is, the professional status of medicine and law come in part from both their class-based elite status and from their control over the organization of knowledge and practices that provide access to capital production. Such elite control of the professions allows capital to be transmitted and inherited through symbolic means such as through cultural capital (the unequally distributed tools for appropriating social valued knowledge, tastes, and dispositions) and social capital (social networks). As Pierre Bourdieu demonstrated, these forms of capital are exchangeable.[5] Capital, for example, can buy access to cultural capital, and cultural capital can provide access to capital. The professions, which serve as the model for teacher preparation reform in the liberal professionalization view, receive their elite status not merely because of the expert control over specific information but by virtue of the professional relationship to both specific information and the forms of capital. Automotive electrical mechanics and plumbers have a high degree of expertise, yet they are not held in high general esteem, as are doctors and lawyers. By claiming political neutrality and the universal value of particular knowledge, the professionalization agenda effaces the specific relationships that particular holders of expert knowledge (professionals) have to the management of the forms of capital (economic capital, cultural capital, and social capital) on behalf of the owners of capital.

The professionalization agenda also effaces the extent to which different economic classes struggle for hegemonic control over expert knowledge. We can benefit here from Gramsci's distinction between traditional intellectuals who learn to produce knowledge for the maintenance of the class order and the organic intellectuals who learn to produce knowledge for the working class.[6] For Gramsci, the professional knowledge producer always produces knowledge specific to class-based political blocs. The point is that the professionalization agenda for teacher preparation reform misleadingly universalizes and depoliticizes what are, in fact, particular approaches to knowledge, which are political. As Lois Weiner rightly argues [and it is worth quoting at length here],

> Although research about effectiveness of teacher education, such as Darling-Hammond's is essential to counter neoliberal claims that teacher education has little value, I suggest that research findings should be joined to a political defense of teacher education's value as a public good. Research funded and publicized by neoliberal think tanks, such as the American Enterprise Institute, is explicitly political, weaving a seamless cloth of neoliberal political assumptions and policies in teacher education that emanate from its principles, especially the way we measure teacher effectiveness. Research about teacher education's value should be rooted in a political argument about our society's need for education as a public good. The decision to surrender the more explicit political rationale for teacher

education's usefulness, as was done with NCATE's decision to jetti-
son the "social justice" disposition in response to conservative and
neoliberal criticisms may seem realistic. In fact, it weakens the case
for public support of institutions that prepare teachers by ceding to
neoliberalism the discourse of educational inequality.[7]

The neoliberal deregulation agenda for teacher education calls
for ending the "monopoly" control that university teacher preparation programs have, bringing in non-educators to the profession,
questioning the value of teacher certification and graduate degrees
in education, and developing more alternative routes to teaching.
In the deregulation agenda, high-stakes tests for teachers would
replace certification programs as the control over entrance into the
profession. Similar to the logic behind the outside leader model for
principals and superintendents discussed in Chapter 4, the deregulation agenda views the state and university as unnecessarily imped-
ing entrance to teaching by educated professionals from other
careers. In this perspective, non-educators should be recruited
from the private sector because they will naturally be infused with
the ethos of "excellence" and "efficiency" which has been, until
recently, closely associated with the corporate world. This assumption accords with the broader neoliberal celebration of business
and the concomitant denigration of all things public and especially
public schools. The deregulation agenda, according to Weiner,
importantly shifts the blame for poor test-based forms of accountability from students to teachers framing teachers as "the problem"
that needs to be overcome to improve the quality of schooling.
While Weiner is correct, we might think of this as *expanding* the
blame from the cultural-deficit accusation levied against the student to both the impugned student and the teacher who are targets
of the neoliberal assault that she decries. The neoliberal deregulation agenda seeks to transform teacher preparation into more of a
"market" with lessened barriers to entry and lessened state control
over standards. The claim is that the only measure of quality of
teacher preparation should be the test scores of students in schools.
Calls for deregulating state and university control over certification are being combined with calls for expanding state and foundation investment in long-term database tracking of student test
performance so that teacher preparation methods can be isolated
based on increased student scores. Venture philanthropists and
most notably the Gates and Broad Foundations push for such

teacher education reforms, and the new direction of the Gates Foundation as of 2008 makes the influencing of teacher behavior and teacher preparation central to the new direction of the domestic agenda.

## The Venture Philanthropy Agenda

In accord largely with the neoliberal deregulation agenda and select aspects of the professionalization agenda, venture philanthropists seek to transform teacher education by funding initiatives for teacher bonus pay that are tied to standardized test scores, facilitating the entry of professionals and especially business professionals into the teacher workforce, fostering and funding alternative certification that is typically of shorter duration and often removed from academic control than traditional certification, funding the implementation and development of database systems for both recruitment of teachers and the measuring of teacher performance in accordance with student standardized test scores, and the funding of so-called school "turnarounds." These "turnarounds" involve firing every employee of a school, smashing the union, and bringing in a business-oriented team to redesign the school and train teachers and leaders while doing so. Both Broad and Gates are involved in multiple examples of these initiatives. For example, both fund "turnarounds," bonus pay initiatives, database projects, and efforts to transform teacher education.

In May of 2009, the Chief Education Officer of the Gates Foundation, Vickie Philips, testified before the U.S. House committee on labor and education and was explicit in the new direction the foundation takes blaming teachers, teacher education and teacher behavior as the central educational problem resulting in low "student success" and the continuation of an "achievement gap." Philips argued that the Gates former emphasis on small schools did not work because essentially the only reform that is worth pursuing is to make what she calls "effective teachers."[8] The "effective teacher" is one whose behavior results in increased test scores. Philips told Congress,

> Effective teachers play the single most important role in accelerating student achievement. The data here are overwhelming. A body of research spanning 30 years has demonstrated that the differences between top quartile and bottom quartile teachers account for vast

differences in student growth—as much as a quarter of the achievement gap per year.[9]

Despite the dubious allegation about 30 years of research (one can find thousands of years of scholarship rejecting the reduction of good education to only teachers), there is a basic problem with Philip's, claims. Her argument is tautological. If "effective teachers" are defined by teachers who increase student test scores, then the statement "effective teachers play the single most important role in accelerating student achievement" really only says "teachers whose students score higher on test are teachers whose students score higher on tests." The circular argument does not tell us much, other than that some teachers have students who do better on standardized tests over time. What it does not sufficiently explain is the value of standardized tests, the value of reducing the quality of the teacher to such tests, the role that teaching-to-the-test plays in increasing test scores, and the toll that has on genuine forms of learning, to name but a few problems.

Philips, a former superintendent of the Portland, OR public schools preaches a kind of "no excuses" demand for test-based academic achievement for poor and minority students. In speaking, she often tells her own bootstraps story of coming from rural poverty and overcoming obstacles through hard work. She believes that differences—racial, ethnic, linguistic, and class difference—are to be largely ignored as the best approaches to teaching and learning and a universal curriculum. Announcing her departure from Portland,

> she said she had a hard time turning down a job that would "touch the lives of millions of young people, proving that all of them, regardless of race, background and income, thrive when given the right education and support."[10]

She was criticized as superintendent for having a similar approach to that of Secretary of Education Arne Duncan when he was "CEO" of Chicago Public Schools: closing schools against the protests of community members, fostering privatization and gentrification, viewing teachers as an obstacle to the change she wanted, and in a relatively progressive city, embracing a view of curriculum that appears to accord less with liberals and progressives and more with conservatives such as E.D. Hirsch with an emphasis on common core knowledge. She was also criticized for failing to raise the test-based academic achievement that she embraces before she left Portland.

> Her foes will point to a pile of unfinished or barely started business: stubbornly poor high school student achievement, a school choice system in need of repair [sic] and a brewing battle between district leaders and teachers over what's taught in classrooms.[11]

Despite Philips's own failure to significantly raise student test scores in Portland and more importantly despite her own dubious perspective on teaching and curriculum, Philips is leading the Gates Foundation in pursuing multiple routes to change teacher education in the direction of the test-based orientation. The solution is the tautology:

> Our considered position is that we cannot narrow the gap or substantially raise performance for all without dramatically increasing the percentage of effective teachers.[12]

Philips explicitly rejects the Darling-Hammond professionalization approach.

> We can't fix this just by recruiting teachers with stronger credentials. After numerous studies, we can say with confidence that master's degrees in education in no way predict which teachers will be effective in the classroom. Likewise high SAT scores, high scores on certification exams, and other impressive credentials fail to predict effective teaching very well. Because we don't know how to predict who will be effective in the classroom and who won't, credentials are a very blunt instrument and will not take us very far.[13]

By defining "effective teaching" exclusively through student test score improvement, the claims about advanced degrees, credentials, and teacher test scores not providing a clear pattern of raising student scores evades a number of basic issues such as in which schools and communities teachers teach, the class, racial, and linguistic differences in the schools and communities, and the impact those have on test scores changes. However, in this view, class and cultural differences are obstacles that need to be registered only for the purposes of identifying an "achievement gap," meaning the differences in test scores but not in developing pedagogical approaches and curriculum that might relate to student experiences, affect motivation, or be the basis for meaningful and even critical forms of learning that relate the subject of study to problems in students' lives and provide the intellectual tools for interpreting and intervening collectively in the broader forces producing those problems and experiences. Schooling in the approach put forward by Philips and Gates belies a "banking education" approach that has no use for such critical considerations. Advanced education study and certification programs do stand as crucial sites where future teachers can explore critical questions beyond test-based methods. The Philips/Gates approach to teacher education aims to reduce teaching to little more than allegedly effective methods for test score improvement gutting out the intellectual, critical, and socially transformative potential of teacher education. Philips announced to Congress that the wealth and clout of the Gates Foundation would be wielded to affect this agenda.

> We are investing heavily in developing measures to determine reliably which teachers are effective and which are not. We are also researching the most promising ways of making the teachers we have more effective. It is essential that we develop and distribute proven mechanisms to improve the effectiveness of teachers. Through several in-depth district partnerships, we will work on realigning policies and practices to better measure and increase the numbers of effective teachers. We will announce those partnerships later this year.[14]

While the liberal professionalization and neoliberal deregulation agendas differ in significant respects, they share an assumption about the possibility of measurable neutral and objective "student achievement," and they assume that influencing teacher practices will change this "student achievement." The neoliberals have a singular focus on "outcomes" which are essentially test-based measures of "student achievement." The liberals frame outcomes less through student test scores than through the professional practice of teachers. Neither the liberal nor the neoliberal perspectives make the "outcome" of teacher preparation the preparation of students for critical intellectual engagement toward the end of collective social transformation.

Part of the allure of the neoliberal deregulation agenda embraced by VP and typified by Gates is how it manages to articulate a number of compelling ideologies at once. First, it appeals through a scientistic promise of objectivity and neutrality—a vigorous embrace

of positivist rationality that separates information from the underlying values and assumptions organizing it. Second, it appeals
through the invocation of the metaphor of private sector efficiency.
Third, it appeals by conjuring market-based deregulation and a
connected animosity to alleged public sector bureaucratic inertia.
Yet, both liberal advocates of professionalization and neoliberal
advocates of deregulation make the mistake of reifying "student
achievement" as an allegedly neutral and objective measure. This
has the effect of denying the inevitable politics of the curriculum,
effaces the power relations tied to particular claims to truth, conceals how forms of authority are legitimated through the securing
of particular representations, separates the underlying assumptions,
values, and ideologies from the organization of information, facts,
and truth claims, and renders difficult to discern crucial public
questions about whose knowledge and views are represented, and
how such representation is tied to broader configurations of economic, political, and cultural power.

Both the professionalization and deregulation approaches mistakenly aim to understand teaching and teacher preparation as apolitical rather than as inevitably political. As Weiner explains, the
neoliberal deregulation agenda is a political agenda; however, its
proponents present it as an apolitical matter of efficacy toward a
goal of economic inclusion. As Cochran-Smith and Fries argue,
claims for reform need to address the underlying ideals, ideologies,
and values behind claims to teacher quality. To take this one step
further, the underlying ideals, ideologies, and values behind claims
to teacher quality ought to be understood in relation to material
and symbolic power struggles over teacher education. In other words,
liberal ideology and neoliberal ideology as applied to teacher
education cannot be examined solely as collections of ideals, but
such ideals need to be understood through the battles waged by
groups for control of schools and school systems, teacher education
institutions and in relation to class struggles which are lived through
particular cultural formations. The making of teachers is not merely
a matter of effective preparation of knowledge delivery agents, but
rather teacher education is a profoundly political undertaking.
As Henry Giroux has argued, there is radically democratic potential
in fostering the development of teachers as transformative
intellectuals.[15]

The imperative for teachers as transformative intellectuals could
not be more pressing than at the present moment in which the

Obama administration attempts to deny politics and instead embraces a post-ideological practicalism. Obama and his pick of Arne Duncan as Secretary of Education repeatedly assert that educational improvement should not be ideologically driven but rather driven by "what works."

Obama spoke directly to this in an interview that was aired on C-Span.

> OBAMA: Well, I think what it means is that I don't approach problems by asking myself, is this a conservative—is there a conservative approach to this or a liberal approach to this, is there a Democratic or Republican approach to this. I come at it and say, what's the way to solve the problem, what's the way to achieve an outcome where the American people have jobs or their health care quality has improved or our schools are producing well-educated workforce of the 21st century. And I am willing to tinker and borrow and steal ideas from just about anybody if I think they might work. And we try to base most of our decisions on what are the facts, what kind of evidence is out there, have programs or policies been thought through. I spend a lot of time sitting with my advisors and just going through a range of options. And if they are only bringing me options that have been dusted off the shelf, that are the usual stale ideas, then a lot of times I ask them, well, what do our critics say, do they have ideas that maybe we haven't thought of.[16]

Obama's declaration of post-ideological practicalism denies his own political agenda while laying out a political agenda. For example, he defines what works for education in the same breath as workforce preparedness—a view of education from the perspective of business. Obama's practicalism evades the questions of "what works to achieve what ends," for whom, why? There are troubling roots to Obama's practicalism in the organizing tradition of Saul Allinsky in Chicago. Allinsky's *Rules for Radicals* makes the case for a project-based kind of politics defined by short-term gains in which proponents of a project should make deals with ideological opponents if it will achieve the ends. There are serious ethical problems in such a view as repayment of the political support often results in organizers later supporting projects that they cannot ethically endorse.

Obama's denial of politics harkens back to the neoliberal post-politics of the Clinton administration and indeed most of Obama's administration is reconstructed from that one. John Podesta, who was Clinton's chief of staff, led the transition team and drew heavily from his own think tank the Center for American Progress that straddles the liberal and neoliberal agendas for education reform.

The post-politics, post-ideology claims from the Obama administration are limited and shortsighted in the ways that they deny power struggles over competing interests and ideologies as well as material resources. For example, the limitations of this view are particularly apparent with regard to the history of racial struggle for educational resources and the role that teachers can play as part of a struggle for racial justice. The post-political view erases the history of the struggles for civil rights, of which schooling was central. As both Obama's administration and the venture philanthropists call for universally valuable knowledge and for "high quality teachers," they dehistoricize the struggles that have been waged for racial justice while also denying how the history of racialized unequal educational resource distribution continues on in the present day. Obama's post-politics shares with the venture philanthropists an embrace of, on the one hand, "evidence" and, on the other hand, a rejection of evidence by calling for "experimentation" with policies that are not based in evidence but rather in ideology. Charter schools and "choice" are an excellent example of this as there has been no good evidence nationally or in Obama's Chicago for either. However, these are politically expedient policy decisions that get political capital for Obama as wide support for this dubious experiment can be found across the two parties. The teacher education approach of Obama/Duncan falls flat, as the empty concept of the "quality teacher" becomes the goal defined by workforce preparation and standardized test scores.

Giroux's radically democratic conception of the teacher as a transformative intellectual differs from that of the liberal one of Darling-Hammond who contends that the justification for the professionalization agenda is equal access to something she refers to as "quality teaching." She also criticizes the deregulation agenda and its overemphasis on test scores as necessarily resulting in rewarding teachers for giving attention to students who show the most likelihood of scoring highly or raising scores and that this will result in the neglect of the students most in need of attention. While she is right about the destructive effects of the overemphasis on test-based outcomes, Darling-Hammond like her neoliberal enemies presumes that "high quality teaching" is static and universal. This differs from the critical pedagogical approach of Giroux, Freire, Apple, Weiner, Sleeter, and others for whom both pedagogy and curriculum are always contextually based. From the critical perspective, struggles against oppression and for emancipation are universally applied. That is, critical forms of education universalize educational projects tied to social transformation projects. For Darling-Hammond, professionalizing teacher education would help to accommodate the otherwise excluded students to an essentially just social order by transmitting the right knowledge to them. The critical approaches, which are transformational rather than accomodationist, prepare teachers who can help students theorize their experiences of oppression to collectively address the systemic and structural causes of that oppression. Whether humanist variation on critical pedagogy such as the Freirean vocation of becoming more fully human, the Gramscian aim at forming organic working-class intellectuals as opposed to the traditional intellectuals who serve the ruling class, or the poststructuralist-influenced variations of critical pedagogy that comprehend both subject and society as constituted by difference—all of these and also critical approaches to teacher education view the teacher as a critical intellectual whose work enacts an emancipatory vision for the future and aims at remaking the institutions and structures of the society in part through fostering critical consciousness and contributing to the conditions for democratic social relations. This means that the teacher, as a transformative intellectual, fosters in students the tools to ruthlessly criticize existing realities and to imagine and build alternative possibilities.

## TRADITIONAL PHILANTHROPIC INVOLVEMENT IN TEACHER EDUCATION

To get a better sense of the professionalization versus deregulation agenda, it is valuable to look at the history of how traditional philanthropy sought to influence education. As Karen Symms Gallagher and Jerry D. Bailey point out, philanthropic foundations have a long history of influencing both medical and legal preparation. Carnegie Foundation was involved in the 1910 Flexner report that dramatically transformed medical education and the 1921 Reed report commissioned by the ABA. In 1994, The Carnegie and Rockefeller foundations funded the National Commission on Teaching and America's Future (NCTAF). NCTAF's report *What Matters Most: Teaching for America's Future* significantly assumed that "what teachers know and can do is the most important influence on what students learn."[17] The principle NCTAF recommendations were:

> Rely on high-quality standards for learning and teaching, reinvent teacher preparation and professional development, recruit qualified teachers for every classroom, encourage and reward knowledge and skill, and recreate schools as learning communities.[18]

Gallagher and Bailey highlight the similarity of *What Matters Most* to the 1910 Flexner report recommendations. They valuably point out that the twentieth-century influence over professional preparation saw a tension between professional expertise and equal participation.

> Democracy requires sufficient education and information for all citizens to enable them to participate meaningfully in decisions affecting them, yet specialized knowledge—expertise—has often stood in opposition to universal and equal participation in public affairs. When the public debate is over what kind of public education is appropriate for all students, both experience and equality of participation are vital to the policies adopted.[19]

Gallagher and Bailey are concerned with the concentrated control over policy formation that philanthropic foundations wield. They point to Andrew Carnegie's emphasis on elevating scientific standards and advancing knowledge aimed at making education a means to popular opportunity as opposed to the Carnegie Foundation's emphasis on influencing governance in exclusionary ways.

> He [Carnegie] had not envisioned the various Carnegie foundations as directly influencing governance because he believed that an educated citizenry diminished the need for any kind of regulatory structure, be it professional or governmental. The reaction of the elite legal community and members of the foundation's board of trustees to the Reed report showed how different Carnegie's commitment to education and its role in a democratic society was from that of the men whom he appointed to the various boards of his philanthropic foundations.[20]

Gallagher and Bailey contend that reform of medical education in 1910 following the Flexner report focused on professional practice while it could have focused on the more democratic concern with public health. They suggest that contemporary crises of healthcare in the United States could be very different today had the reform of professional preparation been at the fore among the concerns of a democratic public.

Gallagher and Bailey draw a parallel to the reform of teacher education and invoke the work of Ellen Lagemann to suggest that the philanthropic foundation stewardship of educational reform resulted in an exclusionary politics of knowledge creation. Gallagher and Bailey suggest that Carnegie had a highly controlled concentration of power over the central questions of the following: the politics of knowledge: "which field of knowledge and which approaches within different fields will be recognized as authoritative and therefore associated with the expertise considered relevant to policy making"[21]; the participation in public affairs: "How should decisions be made and by whom"; the access to the politics of knowledge creation: "who can gain entrance into the elite groups of knowledge producers?" As Gallagher and Bailey point out, Carnegie surrounded himself with likeminded colleagues with similar backgrounds to run the Carnegie Foundation—all of them sharing "interests, educational backgrounds, knowledge bases, and personalities."[22] One might expand this to recognize here a class-based project of professionals shaping knowledge and practices in ways compatible with the ruling interests of a Carnegie, Rockefeller, or Ford.

Gallagher and Bailey are right in emphasizing these critical questions of the politics of knowledge and applying them as a test to the roles played by philanthropies. The same issues with regard to controlling knowledge and also shaping institutions and practices are involved in contemporary VP control over schooling and more specifically teacher preparation.

Gates, Broad, Walton, and the other venture philanthropists largely surround themselves with likeminded advisors and personalities who embrace the basic assumptions and general direction of the foundations. The crucial issue is the extent to which VP is closed to questioning its framing assumptions about the value of teachers and the role of schools in relation to broader public struggles and problems. Put differently, VP in education is insulated from the broader democratic debate, dialogue, and deliberation about education outside the foundations. Consider some of Gallagher and Bailey's questions about knowledge-making in relation to the VP approach to teacher education.

1. "Which field of knowledge and which approaches within different fields will be recognized as authoritative and therefore associated with the expertise considered relevant to policy making?"

In the case of the VP approach to teacher education, the "research" is being drawn heavily and selectively from right-wing scholars and right-wing foundations to make the deregulation case. This is a minoritarian view in education that is being misrepresented by the venture philanthropists as an unquestioned consensus such as in Philips's statement above about "30 years of research."

2. "How should decisions be made and by whom?"

For the venture philanthropists, decisions about reforming teacher education should be made by them rather than by the bulk of educational researchers and policy makers, or by an informed public. The decisions once made by this select and small group of powerful people are to be put into public institutions through "partnerships" with cash-strapped public school districts. The individuals and corporations who benefit from public largess in being able to have these foundations in the first place then use these foundations to steer and influence policy and practice in ways that are at odds with the public interest.

3. "Who can gain entrance into the elite groups of knowledge producers?"

The venture philanthropists are a highly exclusionary group that affords entrance to the super-rich who are aligned with the agenda, those already a part of the project and partially those churning out scholarship sympathetic to it. Those in the public sector and non-profit sector who are desperate for grant money, and who will do whatever is demanded do not enter the elite groups of knowledge producers that dominate the VP agenda.

We might build on these central questions of the politics of knowledge raised by Gallagher and Bailey and emphasize three more crucial concerns: (1) the political economy of knowledge-making institutions, (2) identity formation and (3) distribution and circulation of knowledge. The matters of ownership and control, income and wealth of knowledge-making institutions has much to do with the control that knowledge producers have over the production, distribution, and circulation of knowledge. The VP efforts to transform teacher education has implications for identity formation of both teachers and students as teachers are increasingly encouraged to think of themselves as delivery agents for knowledge produced by others. In this view, teachers are deskilled and deintellectualized. The students are also formed as "educated" by having successfully taken in and regurgitated the official knowledge on the tests in this view. The VP agenda for teacher education compromises the identifications that teachers and students might have with the humanist and critical ideals of the educated person. As well, the VPs, through their enormous resources, are able to distribute and circulate widely the knowledge they produce about what constitutes legitimate and quality teaching and learning. This involves not only heavily publicizing the VP agenda through mass media and scholarly venues, but also by being a prominent financial force that influences others to regulate their own speech. One example of this is how education scholars at research universities are wary of criticizing the VPs for fear of compromising potential grants that have high academic currency.

## VENTURE PHILANTHROPY AND THE GRADUATION AND UNIVERSITY PREPARATION AGENDA

Venture philanthropists have been involved not only in the aforementioned initiatives such as paying teachers and students for grades, creating test-score–oriented databases to influence teacher practices in the direction of standardized test-based "banking education," and funding school "turnarounds," among other initiatives. They increasingly turn toward influencing teacher preparation and K-12 practice toward the end of higher education preparation.

The Gates Foundation has retooled their education agenda toward the end of increasing graduation rates and increasing university readiness.

At the Bill & Melinda Gates Foundation, we believe that every life has equal value and all individuals should have the opportunity to live up to their potential. In the United States, the key to opportunity is education. Education is the great equalizer. It enriches our lives, informs our choices, and prepares us for meaningful employment and to contribute to the communities in which we live. The foundation's work in this country is focused on two major initiatives: ensuring that a high school education results in college readiness and that postsecondary education results in a degree or certificate with genuine economic value. Ultimately, we must ensure all students graduate high school with the skills and knowledge they

need to be successful in higher education...While graduating from
high school is still a milestone worth celebrating, higher education—
whether a two- or four-year college, a technical school, or a certificate program—has emerged as the critical path to opportunity for
students and the avenue to achieve their dreams.

—"Our Mission," from The Bill and Melinda Gates
report *College-Ready*

While the Gates Foundation's emphasis on college readiness is a
worthwhile goal, unfortunately Gates largely measures readiness
through an emphasis on standardized testing. As already described
in the mission statement, the values of higher education are to be
determined by the economic exchangeability of that higher
education degree on the employment market. Obviously, this is a
highly reactive way to understand higher education in which the
needs of corporations dictate what is valuable to learn. As discussed
in prior chapters, one of the crucial limitations of formulating educational value through job opportunities has to do with the structural limitations of the global economy to include educated workers.
As a visit to a taxi stand of one of the many Indian cities reveals, a
highly educated workforce filled with people with graduate degrees
in information technology, computer science, and engineering does
not guarantee employment. It may, in fact, result in a highly educated group of unemployed or underemployed people leaning
against their dormant taxis. Decisions about what kinds of work
should be done and by whom cannot be left strictly to markets.
Students need to be educated for collective organizing, public deliberation, and public struggle to assure better work conditions and
public priorities. Education can better assure these conditions by
offering students the intellectual tools of self-defense. It is specifically this kind of political education that the anti-critical test-based
approach of the VPs prohibits.

Broader and more holistic conceptions of understanding, wisdom, the education of the whole person from the enlightenment
tradition (not to mention criticism of aspects of this tradition) are
diminished in this view in favor of an instrumentalized conception
of knowledge. Additionally, the Gates mission mistakenly understands education as "the great equalizer" when much of what
schooling does, and test-based forms of schooling in particular do
is to confirm and bolster the cultural capital of those students who
come from class and cultural positions of privilege while denigrating

and punishing the cultural capital of materially and culturally oppressed students. For schools to become "equalizers," it would require that they make utterly central to curriculum and school model matters of power, politics, ethics, and history such that claims to truth are taken up in relation to students' experiences. The issue is not that school cannot serve as an "equalizer"; it is that to do so would require a radically different conception of teaching and learning than the ones championed by Gates and the other venture philanthropies.

The insistence of the Gates Foundation on the quality of education being reducible to student test scores, the behavior of individual teachers who can be "incentivized" monetarily and who should be drawn from the stock of professionals, and especially business professionals, renders it diametrically opposed to the progressive value of teachers as transformative public intellectuals. Teacher education as a public and critical project needs to be defended and expanded by those committed to progressive ideals. Moreover, the sites of struggle over the future of teacher education include not only teacher preparation programs themselves but university administration, accreditation bodies, state and national associations, teachers unions, academic and public discourse. The VP approach to teacher preparation (and all aspects of schooling) emerges from a misplaced expansion of market language and logic to all realms of social life—that is, to economism. The next chapter aims to explain and question some of the most basic assumptions of this economism that animates VP and propose an alternative way to think about educational obligation beyond the restricted logic of the market.

# The Gift of Education:
## Education Beyond
## Economism

### Educational Economism Right and Left

This chapter confronts the ways that education has been largely overtaken by economic language and logic, and it suggests an alternative. It begins by criticizing educational economism across the political spectrum and then draws on the work of Marcel Mauss to propose rethinking of educational provision and obligation. The fiscal right has championed economism in the form of neoclassical economics or neoliberalism. This is the dominant economism applied to education, and it drives educational policy, curriculum reform, school practice and culture, and pedagogical ideals while redefining public education through the lens of private business. As part of a broader social logic, right-wing economism applies business ideals and categories to schooling by using such terms as competition, choice, efficiency, consumption, and accountability to describe educational policy, and in so doing, shifts the public and democratic possibilities of schooling to the realm of the private and the for-profit. Neoliberal education applies a managerialist logic of rationalization, imagining schooling as a business enterprise, describing public education as a private market, thinking of knowledge as commodities, and proposing an entrepreneurial vision for teachers who are expected to compete for grants and for merit pay, and for students who are offered money for grades.

Although the neoliberal agenda that celebrates privatization and deregulation has come into crisis arguably since the late 1990s, it became particularly prevalent in public discourse with the financial crisis of 2008 that resulted in the need for the federal government of the United States to rescue multiple industries including financial and automotive and even nationalizing major industries and corporations. In this crisis, a central tenet of neoliberalism—that markets can regulate themselves—has been recognized in public discourse as dubious at best. Even the most hardcore proponents of deregulation such as Alan Greenspan have admitted that such assumptions are wrong. However, neoliberal educational projects will not simply end by themselves. The values and organization of the industrial production economy continue in schools in the form of the Gary Plan, Taylorism, scientific management, and the corporate model of administration, to name a few. Despite the fact that the United States has largely shifted from an industrial to a consumption and financial economy, the public school system remains mired in the trappings of industrial economism. These vestiges of earlier economic ideals, which were fought for by groups such as the National Association of Manufacturers and foundations, continue in educational bureaucracies, perpetuated by institutional practice and culture, ideology, and perhaps just bad habit. Just because neoliberalism as an economic doctrine may be in question or even fully discredited does not mean that neoliberal ideals will stop being pushed in public education.

Neoliberal education drives a privatization agenda that includes running public schools for profit, voucher schemes, charter schools, for-profit contracting, and commercialism. As an economic project, this treatment of public schooling tends to redistribute public wealth and educational resources upward. The political and cultural costs of such economic and symbolic shifts are great. The social, political, and cultural dimensions of education are impoverished in this view, as market considerations reign supreme. In the accomodationist view of neoliberalism, the only salvation for the individual in a radically unjust social order is to fit in or perish. Education is the means through which the individual is to fight to the top of the heap. The social possibilities of neoliberalism are framed largely through national economic competition in the global economy. The economy is naturalized as the best and only way to think about education among other social issues. Unfortunately, economism has also resurged with the education left.

Marxist educational theories have offered crucial insights about schooling in capitalism. Some of these insights include how capitalism configures knowledge as a commodity, transforms schooling on the model of the industrial economy, and reproduces the social conditions for the reproduction of capital by teaching skill and know-how in ways ideologically compatible with capitalist social relations and class position. As well, such Gramscian Marxist social theorists as Althusser[1] have offered a theory of subject formation: interpellation. Gramsci's thought on the relationship between politics and education has inspired a number of critical pedagogues to theorize the role that schooling can play in hegemonic struggle.[2] Other Marxist thinkers on education such as Raymond Williams inspired critical engagement with the question of the "selective tradition," questions of canonicity which open up critical concerns about the production, distribution, and reception of knowledge, its relations to the securing of social authority, and its relationships to broader structures of power.

Yet, Marxist economic reductionism has made a disturbing return to the field. While unlike neoliberalism this camp has the virtue of focusing squarely on oppression, social justice, and class warfare, the new old champions of Marxism have not bothered to learn from the many decades of criticism inside and outside the field of education.[3] The new old Marxists embrace an anti-democratic vanguardism and class reductionism while failing to deal with the many problems of the Marxist legacy of thought, including the anthropocentric tendency to view nature as ideally exploitable for human uses, the reduction of human value to labor, the patriarchal and racist legacies of the Marxist inheritance, the theory of culture and ideology in Marxism that reduces both to reflections of the economic base, the teleological theory of history, the mechanistic theory of agency determined by class position, the modernist enlightenment tendency toward purity, unity, totality, and harmony that is contrary to thinking difference, among other problems. There continue to be crucial insights to be taken from Marx and the Marxist tradition of thought. However, this "purist" Marxist revival remains trapped in a theoretical time warp, and, more importantly for this book, it remains stuck in what Baudrillard referred to as "the mirror of production"—that is, the expansion of the productivist metaphor throughout social life, the extent to which Marxism remains trapped within the assumptions of the political economy and metaphysics that inspired it, and the restricted understanding of human life and activity by labor.[4] Like neoliberalism, its nemesis, Marxist education is hopelessly bound to economism. Critical educators such as Henry Giroux and Stanley Aronowitz have pointed this out since the 1980s.[5]

Both neoliberal education and Marxist education, due to their economism, suffer from difficulty theorizing the distinction between public and private, the politics of knowledge, and the power-infused workings of culture. Both advance a vision of education inherently incompatible with the democratic possibilities of public education. Neoliberal education collapses the public possibilities of public schooling into the private possibilities of amassing wealth. The public roles of shared control and the civic possibilities of public deliberation, debate, and dissent are undermined by neoliberalism's push to privatize public schooling, treat students as knowledge consumers, and treat teachers as deliverers of commodity. The value of knowledge is reduced in the neoliberal view to its exchangeability in the marketplace. Struggles over claims to truth and the relationship between knowledge and power are largely ignored as the neoliberal view pushes hard for the standardization of the knowledge deemed universally of value. Neoliberal education imagines the future as an endless present. Within the supreme imperative for continued economic growth, the possibilities for the individual reside with the uses of education to give the educational "consumer" an edge at competition, and national policy should, in this view, be based on global economic competition. The vicissitudes of the market, the economic exclusion and impoverishment of roughly one-half the planet, and also the crises of value and political cynicism produced by the ascendancy of consumer culture merit hopeless shrugs by the neoliberals whose motto might as well be "adapt or die." As Lawrence Grossberg suggests, neoliberals are faced with a contradiction when it comes to ethics and morality. While, in principle, the market is supposed to be self-regulating and to produce the best social outcomes, few neoliberals would allow absolutely everything to be made into a market—human organs, for example—and so, they must legitimate their policies through reference to other discourses for morality, such as American exceptionalism or Christianity.[6]

The vulgar variety of Marxist education fails to appreciate the public/private distinction by treating public schooling as little more than an arm of capital and a tool of capitalist oppression. Antonio Gramsci stands as a glaring exception to this, having profoundly theorized the struggle for civil society as central to the making of a hegemony forged through consent.[7] For Gramsci, the "private" realm of civil society must be won by ruling groups, and it is not only the "public" realm of the state and its juridical and coercive institutions that must be seized to transform the material relations of production and hence the consciousness of men. Consequently, the struggle for ideas, the actions of the intellectuals including teachers are not just profoundly political but they are crucial to the winning of the social order by competing classes. Politics is an educational project, and education is necessarily political. The role of public schooling as a public democratic institution that prepares citizens for civic engagement is not central to the new old Marxian education any more than it is to neoliberal education because the Marxists imagine a post-revolutionary future that will be run as a dictatorship of the working class. For the Marxists as for the neo-liberals, democracy receives lip service, but these unlikely bedfellows share a commitment to reducing education to economics. While the Marxists rightly attack the damaging structure of global capitalism and its human costs, they are left with no place to go as the alternative educational form can only be derived from class analysis. The crucial questions of cultural politics and the rejection of the theoretical tools developed by multiple traditions of cultural analysis about whose knowledge should matter, what knowledge should matter, and how language, meanings, and ideologies relate to material struggles leave such Marxists stunted.

The new old educational Marxists view every other perspective and insight about education as a threat to the one true cause of class. Feminism, racial justice, progressivism, socialism, postmodernism, poststructural theory, liberation theology, postcolonial theory, cosmopolitanism, and the legacy of philosophical liberalism from which all of these derive all appear to the educational Marxists as a threat to be annihilated. As Dave Hill, a leader of this perspective, announced at the annual American Educational Research Association conference in 2007:

> Non-Marxist and Anti-Marxist political forces fail to recognize and combat the essentially class-based oppressive nature of Neo-Liberal Capital. Such forces include Extreme Right Racist/Fascist, Extreme Right Populist, Conservative neo-liberal, Neo-conservative, Third Way/Revised Social Democratic (e.g. Die Neue Mitte/New Labour), Christian Democratic, centre-Left Social Democratic, and religious fundamentalist movements and parties, whether they be Islamic, Christian, Jewish Hindu or other religions. .... Objectively, whatever our race or gender or sexuality or ability, whatever the individual and group history and fear of oppression and attack, the fundamental form of oppression in capitalism is class oppression.[8]

In the new old Marxist discourse of purity, different views appear as a danger to the one big truth.[9] Such a perspective is hostile to debate and deliberation and tends toward political fundamentalism rather than the kind of public agonism necessary to democratic culture and governance. As political theorist Chantal Mouffe writes,

> While antagonism is a we/they relation in which the two sides are enemies who do not share any common ground, agonism is a we/they relation where the conflicting parties, although acknowledging that there is no rational solution to their conflict, nevertheless recognize the legitimacy of their opponents. They are 'adversaries' not enemies [to be destroyed]. This means that, while in conflict, they see themselves as belonging to the same political association, as sharing a common symbolic space within which the conflict takes place. We could say that the task of democracy is to transform antagonism into agonism.[10]

We can add here that the critical possibilities of public schooling likewise foster democratic culture by both recognizing the inevitable antagonism at the core of the social but also by teaching the theoretical and political tools for hegemonic struggle.

Neoliberal education is likewise authoritarian in its active denial of politics in favor of the magic of the market. Neoliberal education is fundamentalist in two ways: it is a manifestation of market fundamentalism while denying that there is a politics to the managerial role of markets.[11] The neoliberal perspective wrongly insists that free markets govern democratically as people vote with their dollars. What the neoliberal view misses altogether is how the economy functions politically to position people hierarchically based on their capacities to act in the market—capacities to act which are hardly equally distributed.[12]

## THINKING EDUCATION BEYOND ECONOMISM OR A DIFFERENT KIND

The central question of this book is: what is the alternative to this economism in education that is typified by VP but found across the political spectrum? On one level, the possibility of an alternative to economism could suggest the possibility of escape from the logic of the market. In education, many writers from a critical perspective, I included, have criticized commercialism, corporatization, privatization, marketization, and managerialism by relying on a correct

yet somewhat overly neat distinction between politics and economics. For example, school commercialism treats public schools like
private space, treats citizens like consumers, and treats service in the
public interest like the private pursuit of profit. While this is true,
and public democracy is undermined by school commercialism,
there are a number of issues that get glossed over in such a treatment that posits politics and economics as mutually exclusive. What
can be lost is the recognition that the economics of education is
inherently political in that economics presupposes particular forms
of social organization and hierarchy, public and private values and
priorities, and methods of achieving them. What can also be lost is
the recognition that the politics of education cannot be understood
as wholly noneconomic. For example, liberal criticisms of school
commercialism most commonly argue that turning students into a
captive audience for marketers threatens the innocent space of the
school where disinterested knowledge can be taught. This framing
misses the inevitable class-based assumptions undergirding particular framings of truth claims, the unequal distribution of cultural
capital and the mechanisms such as testing that simultaneously
accomplish and conceal this at the same time, and more broadly the
extent to which different classes struggle to capture public schools
to foster particular class agendas and implement class-based ideologies, to name but a few examples. This is not to say, as neoliberals
do, that politics should be replaced with economics or that, the
magic of the market is inherently democratic. However, the politics
of education are played out in monetary and non-monetary systems
of exchange: material exchange, symbolic exchange, and social
exchange.

Some of the central questions that stretch deeply into the politics
and economics of education involve *how social obligations are forged*
and *how a society recreates itself through the practices that comprise
systems of exchange*. This is not only a matter of formal schooling.
Rather, the political and economic practices that comprise systems
of exchange to create society are pedagogical in the sense that
people learn these practices and teach them to others. The pedagogical aspects of politics and economics offer crucial points of
intervention for individual and collective praxis[13]. How do educational theory and practice foster in a society more democratic or
more authoritarian social relations? What, for example, lies behind
the inducements people experience to form public versus private
identifications, as well as equalitarian versus hierarchical, collective

versus individualistic kinds of sociality? How are social obligations forged through multiple forms of exchange in particular ways that cohere with democratic and just ideals? These questions also open up some new directions for developing critical pedagogies in terms of motivation and desire. But, these questions also reveal that much more is at stake than the future of public education beyond economism. If public education is implicated in the recreation of society, then the broader matter of the future of public education really concerns the future of society itself and the ways that the political system, the economic system, and the culture can become more democratic, just, and equal in part through the forms that schooling takes.

## TOWARD A GENERAL ECONOMY OF EDUCATION

Educational economism, as I have discussed it so far, refers to the framing of educational issues, practices, and policies through restricted or scarcity-based economics. However, there is an under-explored tradition of thought that allows us to imagine education beyond scarcity-based economism. The departure from this kind of restricted economy comes not from imagining a completely non-economic model but rather by expanding economy to include all aspects of life. Referred to as "general economy," this approach takes as its starting point certain assumptions about human beings that differ from the assumptions of restricted economy. While restricted economy treats the individual as an autonomous, calculating, rationally directed and self-interested actor, general economy treats the individual as collectively engaged or socially interconnected. The motivations of the individual can never be understood apart from the broader social meanings and values within which exchanges take place. General economies share a tendency to challenge restricted economy by expanding economic questions to all social life; yet general economic approaches differ significantly. In what follows, I will consider one version of general economy and discuss their implications for a general economy of education. It is important to mention that these views on general economy follow a historical trajectory and yet are in many ways philosophically incompatible.

In what follows, I discuss the general economy of Marcel Mauss. Rather than seeking to offer an exhaustive account of his theory, I focus specifically on his departure from scarcity-based economics. Because scarcity-based economics has overtaken education, this project attempts to plumb this theory for the educational implications of thinking economics and social organization differently.

## Mauss

Marcel Mauss was a nephew and student of Emile Durkheim. His work includes sociological and anthropological study of the history of contracts, sacrifice, and magic. His most influential work was *The Gift: The Form and Reason for Exchange in Archaic Societies*[14] (originally published *Essai sur le Don* (1922)). *The Gift* is considered a founding work of economic anthropology and had an enormous influence on ethnographic methodology. In the 1980s and 1990s, a number of anthropologists engaged with Mauss's ideas on gift exchange, and the early 1990s saw a renewed interest in Mauss in a number of disciplines in the humanities and social sciences including literary theory, philosophy, feminism, and sociology. Aside from the influence Mauss had on the important sociological and education work of Pierre Bourdieu, little has been written in the field of education that considers the significance of Mauss's theory to address economism in education and its manifestations in the form of privatization and managerialism or the rise of philanthropic giving and foundations in education.

In *The Gift*, Mauss studies the gift exchange practices of several "archaic" societies—particularly, Native American, Polynesian and Melanesian peoples, as well as the ancient juridical origins of similar principles of exchange that he traces from Rome, India, and ancient Germany to the present. Mauss seeks to establish the universality of exchange in all societies while distinguishing the different systems of exchange that govern different societies. Mauss challenges the idea that gift exchange in so-called archaic societies could possibly involve the giving of unreciprocated gifts. Such competitions for giving as found in the Native American potlatch (festivals of giving) or the South Pacific Kula (Pacific Island trade of traditional bead jewelry) do not indicate an absence of market exchange principles. Rather, these are social forms of reciprocity that bind a society together through the gift-giving practices. There is, for Mauss, no such thing as a "free gift." This is so because giving a gift creates an obligation within a system of exchange.

All societies for Mauss are reproduced through exchange practices. Collectively, gift- giving forges social obligations and recreates the society through transactions and associated rituals. A central dimension to Mauss's analysis is the distinction between contemporary market systems with their restricted meaning of exchange and what he calls the "total system in services" or "prestation" found in so-called archaic market systems. These archaic economies involve exchange processes and objects of exchange that are meaningful in relation to the whole rest of the society: religion, spirituality, familial and tribal relations, material production, and consumption. Exchanges of goods in archaic societies were not individually but rather were collectively conducted, and their importance was not merely economic. The potlatch ceremony of the Pacific Northwest involved competition by chiefs for giving away or destroying vast wealth. Woven goods, salmon, or metal objects, would be given away to visiting tribes, and the status of the chief would be reflected in the capacity for giving. The capacity for giving established the status of chiefs in relation to others.

Mauss's innovation was to recognize that potlatch was not a noneconomic activity as had been believed before him. Potlatch is strikingly dissimilar from contemporary U.S. capitalism in that individual status in potlatch comes not primarily from the accumulation of wealth, the keeping of wealth, and the conspicuous display of kept wealth, but rather primarily from the dissipation of wealth, the giving of wealth, the spending of wealth, the squandering of wealth, and the conspicuous display of such acts of expenditure. What is also striking is that such transactions are tied to other familial rituals, tribal rituals and relations, religious practices, and moral values.

When you walk into a convenience store and plunk down a handful of change for a drink, the transaction does not call into play your extended familial relations, evoke your moral values or cosmology, or mark your status or that of the cashier in a way that has significance to the entire community. Rather, the exchange act in restricted economic systems is alienated from those other social relations and considerations.[15]

For Mauss, the total system in services or prestation found in archaic economies is a developmental antecedent to gift exchange (potlatch or kula) and an antecedent to, what is for Mauss, the most developed system, capitalism. In systems of gift exchange, reciprocity functions through the ongoing production of obligation. A central question then for Mauss to answer is: why does giving create an obligation?

> If one gives things and returns them, it is because one is giving and returning 'respects'—we still say 'courtesies'. Yet, it is also because by giving one is giving *oneself*, and if one gives *oneself*, it is because one 'owes' *oneself*—one's person and one's goods—to others. (Mauss, 46) (emphasis in original)

Cycles of owing things and owing selves replicate and perpetuate a society for Mauss. Mauss locates this production of obligation in societies practicing gift exchange, and he traces it back through ancient Germanic law and morality, and forward to contemporary morality embedded in the economy. In the gift exchange of potlatch and kula, when a "big man," a chief, gives, the gift must be returned, or it threatens to wound the recipient who does not give back. This too Mauss sees as carried forward to capitalism, "...the whole tendency of our morality is to strive to do away with the unconscious and injurious patronage of the rich almsgiver" (Mauss, 65). The poor person who receives the alms owes an obligation back to the rich almsgiver.

Mauss was writing in pre–War France, after the enactment of vast social insurance legislation. A socialist, Mauss explains welfare state provisions in terms of social reciprocity. He explains that the worker gives not only his labor and the products of his labor in his work but his *very life* to the employer and to the collective. The society in the form of the state and the employers owe the worker security from joblessness, illness, infirmity, and death. Of course, in the present day United States, the political right argues against social insurance programs, indeed the entire series of social programs enacted since FDR, on the basis that these are injurious gifts given by the society to the undeserving poor who are hurt psychically by being unable to reciprocate this generosity. Welfare, housing assistance, and medical assistance do not help the recipient, in the logic of the political right, because they reduce the individual's self-esteem and self-reliance. Such thinking dominates not only conservative politics in the United States but was expressed perhaps with greatest enthusiasm by the neoliberal policies of the Clinton administration. Welfare was replaced with workfare on the assumption that individuals, in order to be deserving of state care, need to prove themselves useful and productive in a restricted economic sense. As policy, workfare clearly expresses the opposite of Mauss's views of obligation. If one has worked for years, given one's life, serving an employer to the financial benefit of the employer and the benefit of society, the employer and the state has no obligation to repay the individual's sacrifice with security. Instead, the individual's continued existence and very life must be justified to the state and the private sector through productivity. Workfare defies unionization and often pays lower than the minimum wage, so for those desperate for a living wage, workfare actually repays the individual for past sacrifice with punishing conditions.

Mauss's perspective (and we can add that of liberal welfare state thought and market socialism) begins with the assumption that the well-being of an individual is in part the responsibility of the whole society. The dominant neoliberal view posits an individual whose well-being is radically the responsibility of the individual or the most private social unit, the family. The initial gift, for Mauss, comes from the individual who works and contributes to the society. For Mauss, the gift begins with the willingness of the worker to, in a sense, sacrifice his life by selling it to the employer. For Mauss, the sacrifice of one's life for labor results in the production of products that contribute to the entire society. On this, Mauss missed a fact that Marx is especially clear about. Namely, the product of work does not predominantly benefit the whole society but rather only those who receive a profit from the surplus value to the extent that they can buy the product. As to the supposed social good of job creation, that too depends on whether the job in question is a good one or a bad one—something that has much to do with struggles to organize labor. For the neoliberal, the initial gift comes from the employer's willingness to profit from the worker's willingness to sell his labor power.

## THE ROLE OF EDUCATION IN MAUSS'S THOUGHT

Education plays a distinct role in Mauss's thought. Mauss views the history of humanity until the development of law and late economic systems as having been subject to violent antagonism. He views peaceful and stable societies as having *learned* to give, receive, and give in return. That is, peaceful and stable societies must learn opposition and exchange that does not aim at annihilating the other group or the other person. Two groups can fight, or they can negotiate. The latter must be *learned*.

To trade, the first condition was to be able to lay aside the spear. From then onwards, they [societies] succeeded in exchanging goods and persons, no longer only between clans, but between tribes and nations, and, above all, between individuals. Only then did people learn how to create mutual interests, giving mutual satisfaction, and in the end, to defend them without having to resort to arms. Thus, the clan, the tribe, and peoples have learnt how to oppose and to give to one another without sacrificing themselves to one another. This is what tomorrow, in our so-called civilized world, classes, nations, and individuals also must learn. This is one of the enduring secrets of their wisdom and solidarity. (Mauss, 82)

Though Mauss's politics, particularly at the end of *The Gift*, have been criticized as being naïve,[16] Mauss's politics shares with contemporary political theorist Chantal Mouffe this central distinction between agonism and antagonism. However, while Mouffe, in the critical tradition of radical democracy emphasizes agonism, the inevitability of conflict, the constitutive incommensurability at the core of the social, deliberation, and difference, Mauss, unfortunately, idealizes an overcoming of agonism in the collective sharing of the common wealth. Nonetheless, Mauss's vision of peaceful shared material abundance relies on education.

Peoples, social classes, families, and individuals will be able to grow rich, and will only be happy when they have learnt to sit down, like the knights, [of King Arthur's round table] around the common store of wealth. It is useless to seek goodness and happiness in distant places. It is there already, in peace that has been imposed, in well-organized work, alternately in common and separately, in wealth amassed and then redistributed, in the mutual respect and reciprocating generosity that is taught by education. (Mauss, 83)

Education for Mauss is both central to the establishment of the good life that overcomes violent conflict, and it is by its nature a model for human interaction. For Mauss, the education process is one of mutually respectful exchange. Mauss's view on the centrality of education to the ongoing making of the good society and the public good brings it into proximity with numerous educational perspectives that make education central to political life, from Aristotle to Deweyan reconstruction and critical pedagogy.

More than this, what is particularly significant about Mauss's view on education in relation to his theory of gift exchange is how it relates to both dialogic theories of education and contemporary poststructural theories of culture that have had an impact on critical education, such as the representational theory of Stuart Hall. Stuart Hall's theory of culture centers on linguistic exchange to explain how meaning is produced, consumed, and distributed, and how identities are forged through the always unequal exchange of signifying practices that are both symbolic and material. All of these perspectives share recognition that education has a direct relation to the rest of the society. Economism in the restricted sense compartmentalizes education, knowledge, and curriculum and renders the crucial political and pedagogical relationships between knowledge and authority, identity, and social formations difficult to discern. Thinking of education in terms of prestation or the total system in services emphasizes the connections between education and the broader society, and it emphasizes multiple exchanges that play out through educational processes.

Economistic forms of public education aim to serve the student's *economic interests* by preparing for their future of work after the schooling process. Mauss reminds readers in *The Gift* that the language of individual economic "interest" is historically recent, owing to the victories of rationalism and mercantilism. "One can almost date—since Mandeville's The Fable of the Bees—the triumph of the notion of individual interest."(Mauss, 76) Mauss criticizes utilitarianism that aims to reduce human goods and desires to economic calculation. He describes "homo oeconomicus" as "icy" and "a machine" though he recognizes the particularly middle-class propensity for icy calculation.

> It is perhaps good that there are other means of spending or exchanging than pure expenditure. In our view, however, it is not in the calculation of individual needs that the method for an optimum economy is to be found. I believe that we must remain something other than pure financial experts, even in so far as we wish to increase our own wealth, whilst becoming better accountants and managers. The brutish pursuit of individual ends is harmful to the ends and the peace of all, to the rhythm of their work and joys—and rebounds on the individual himself. (Mauss, 77)

Mauss counters the individualism of the economic person with a vision of a social person. He anticipates mass desire for a society characterized by the following: a belief and trust in society and the occupational grouping; the individual giving of cooperation to society as being understood as in the individual interest combined with a distrust of individual greed; for the generous giving to society and the occupational grouping, both will repay the gift in double and forgive the individual (Mauss, 77). Mauss offers a strong sense of the reciprocal relation between the individual and the public. While he views social relations economically, he does not reduce social responsibility for the individual to economics the way that neoliberalism does. The emphasis of Mauss on the relations between the individual and the rest of society has significant implications for recognizing the limitations of the neoliberal approach to education that delinks education from its broader social implications.

The capitalist logic of alienation plays out in education to treat schooling as a consumable commodity and treat knowledge as units of product to be consumed by students. As privatization and managerialism become more acceptable as school reform, information becomes more and more a saleable commodity the value of which is decreasingly social and increasingly economic. The estrangement described by Marx with regard to life in capitalism increasingly applies to education. Increasingly, the value of an academic subject is its economic exchange value rather than its social value. This appears in the United States in such trends as the state investment in science and math concomitantly with the eradication of subjects oriented toward less specific market applications—music, art instruction, and physical education. Those subjects that appear to be about living a good life are ended because economic rationalization makes all goods for sale and for consumption. Marx writes of the estranging role of money in the *Economic and Philosophical Manuscripts of 1844*, "I am *stupid*, but money is the *real mind* of all things and how then should its possessor be stupid? Besides, he can buy talented people for himself, and is he who has power over the talented not more talented than the talented? Do not I, who thanks to money am capable of all that the human heart longs for, possess all human capacities? Does not my money therefore transform all my incapacities into their contrary?"(103–104) As education becomes more and more commodified, stupidity appears as its opposite, and the casualties are curiosity, investigation, disagreement, debate, dialogue, deliberation, and dissent—habits of mind which are not only intellectual assets but crucial qualities for the participation and ongoing recreation of democratic culture.[17]

## Extrapolating from Mauss:
## Toward a Theory of the
## Educational Gift

Mauss's analysis lends itself to insights about educational econo-mism. First, his justification for state-organized social provision grounded in reciprocal obligation can be extended to include public education. Public education in the United States, as a result of economism, has increasingly inverted the social obligation to pro-vide education. For neoliberalism, there is a twofold obligation that the individual owes to the society for the gift of public education: the individual's education is expected to contribute to the national effort in global economic competition, and the individual is respon-sible for optimizing the educational system toward the end of upward individual economic mobility. In both cases, for neoliberal-ism, the capitalist economy is the founding gift given to the indi-vidual by the society. The individual repays the gift by successfully navigating first the education system and then the economy so that the individual is served with job opportunities and the nation is served by beating other countries in the global competition for wealth. Nothing illustrates this neoliberal sense of obligation better than the endlessly repeated mantra found across educational schol-arship and discourse about the need to educate students to become "productive" members of society.

We might extrapolate from Mauss's discussion of other social services a different way of thinking about the obligation relations between the public sector and the individual. The founding act of public schooling is not the willingness of the public school system to give education to the student. Rather, public school education is owed to the student for the student's willingness to participate in compulsory public schooling in the first place. That is, just as Mauss explains that the worker gives his life to the employer to labor for the production of goods that contribute to the society, we might consider that the student gives his or her life to the public sector to labor in school for knowledge and dispositions that will in time provide the economy with workers and provide the public sector with citizens. These workers and citizens, in a democracy, are ide-ally responsible to return the favor by participating in self-gover-nance and can expect to have that contribution returned with public respect for rights and provision of security. If the public school system owes the student for his or her sacrifice of their time, freedom, and life, then certain educational realities in the United States appear as a particularly stunning betrayal of the "educational contract." The essential warehousing of poor and working-class students in public schools that are denied adequate investment and resources is a bad bid, a "low-ball offer" by the whole society to the student. The student essentially says," I pledge my life, I give my body to the society when I show up at school." The well maintained, staffed, and funded school that fosters the student's intellect and imagination is more than a *statement* that says, "Student, your growth and development and life is valuable to us." It is a *material* exchange. It is a material and symbolic counter-offer to the pledge the student makes of his or her life. As such, the condition of the schooling offered gives students information as to their social worth according to the society. The dilapidated school building with broken windows and rats, without textbooks, with too few teachers, is more than a denigrating statement that says, "student, this is what your life is worth to us." It is a material exchange. The student offers his young body and ready mind. The school and its contents are a counter gift.

Considering schooling in terms of gift exchange opens up a new way of thinking about such questions as student attendance, discipline issues, student resistance, and public school privatization. For example, for-profit educational management organizations (EMOs), like The Edison Schools, have continued to expand in large part because the poor and working class families whom they target have been historically shortchanged by public schools that have received dismal investment and commitment. Despite the fact that EMOs skim off public tax money, are plagued by testing and financial scandals, and have high teacher turnover due to anti-union policies among other problems, EMOs still appeal to many parents who feel betrayed by the public system. Specifically, if a public school says to a student, "you are worth a broken building without books," then the student might likely respond, "I am going on the market for a different school." This is precisely the goal that educational privatization advocates want to achieve. Even though there is no evidence that privatization results in better schooling (traditionally defined), for-profit companies are able to appeal to potential families by adopting the sheen of professionalism and creating an image of corporate culture and hence an association with wealth and investment: new paint, uniforms, fancy websites, and so on. Even if EMOs do not invest more per pupil in a school than the public system, they are often savvy about looking like they are through such image manipulation even when they are actually skimming public resources. Public school advocates need to realize that the failure to provide adequately for all public schools threatens to privatize not just the poor schools in poor communities that are currently being targeted but all schools in all communities. The failure of the public to deliver universally on the educational contract results in the privatization of exchange relations. In other words, if individuals pledge themselves to the society and the society does not respond in kind, then it should come as no surprise when the individuals opt out of the social contract. This can take the form of turning to privatization, but it can also take the form of dropping out of school altogether or staying in school but in an oppositional defiance to the education offered.

Rethinking educational policy in relation to material relations of reciprocal exchange offers a way of criticizing NCLB that has not been prevalent in much of the mainstream or scholarly coverage. NCLB defines educational quality through numerically quantifiable forms of accountability that it aims to enforce through punitive measures. The managerial logic of NCLB demands demonstrated acquisition of official knowledge before paying out much needed federal resources to schools. It thereby overturns the long-standing basic assumption about public schooling that it is publicly provided by the society to the individual. Holding resources and schools hostage to conditional "benchmarks" of "progress" appeals to a segment of the population that presumes that students and teachers want to "get away with" not mastering the knowledge deemed worthy of knowing. So, the law has to "enforce" learning with the threat of punishment. The punishment potentially means dissolving or privatizing the school as well as pushing through a managerialist charter school agenda and for-profit tutoring in the form of Special Educational Service providers (SES).

NCLB makes what should be working on the logic of the gift into the logic of the hostage. Students pledge themselves to the social by their participation in the system and the society responds with a statement of distrust and a gesture of cheapness. In other words, individuals pledge themselves to the social without reserve, and the social responds with a conditional pledge. To offer an analogy with another public institution, this would be like the U.S. military receiving new recruits and then telling those recruits once they arrive on the battlefield that they will need to prove their mettle in battle before receiving their rifles, tanks, artillery, and air support in order to get the rest of the weapons necessary to fight. Of course, aside from the issue of losing the battle, this would appear as a betrayal by the society because the conditionality on the social commitment would not only likely result in death and disfigurement, it would also mean that those who were getting the benefit refused to honor the pledge made by the recruit with the expected counter pledge. Soldiers in such a predicament would not consider themselves a soldier but rather a hostage. (In fact, the same is true when the soldier is armed but ordered to face battle without a clear mission or competent leadership). The student, who is told that the resources will follow the performance, is much like the soldier betrayed and abandoned. While death may be less immediately imminent for the student, there are nonetheless real material stakes for the student.

For Mauss, the individual impulse toward giving is situated in a particular social configuration. The individual does not have a natural impulse or inclination toward social forms of giving that are transcultural and transhistorical. Rather, within the particular systems of exchange, individuals are educated into dispositions of giving. In other words, the problem with strict economism in education and throughout the society from Mauss's perspective does not have to do with it being unnatural for the individual or at odds with human nature or instinct. The individual has no more of a natural disposition toward giving than the individual has a natural disposition toward violence. There are, for Mauss, social costs to organizing societies through strict economy. These costs include the diminishment of the richness of individual experience as everything becomes subject to utilitarian calculation. However, we can take Mauss's thought a step further and realize that the public must be educated into either social obligation *or* a morality of individual utility and productivity. For economism to be accomplished, the public must be educated *out of* the historically vestigial forms of social obligation (coming from religion, familial values, civic virtues) and into a morality of individual utility and productivity. Economistic "pedagogies" like neoliberal ideology and the venture philanthropy that champions it, result in the building of social formations that tend not toward peaceful agonistic exchange and the fostering of the common wealth but rather of social formations that tend toward violence, antagonism, and distrust. Such social formations in turn tend to educate individuals in individualized, economistic antisocial ways informed by distrust, insecurity, and stinginess.[18]

## CONCLUSION

We can appropriate from Mauss and other theorists of general economy (such as Bataille, Bourdieu, and Baudrillard who deserve attention elsewhere) crucial insights about contemporary educational thought. From Mauss we can understand public educational provision through the relations of obligation and material and symbolic exchange. We can recognize that there are effective "educational contracts" being forged by educational institutions, and we can understand that the public democratic value on democratic schooling and democratic culture requires that the society fulfill its "end of the bargain" with the student both materially and symbolically. We also get from Mauss a recognition that the process of schooling cannot be thought of as discreet and disconnected from what goes on throughout the rest of the society. On the contrary, schooling, as one institution central to recreating society, has to be understood as fraught with political struggles and ought to engage social realities outside of the school, including how the political, economic, and cultural systems work and ought to offer students the theoretical tools to interpret and act on these interpretations, to challenge and change forces of oppression.

We can also see the limitations of the contemporary utility-mindedness, the hyper-rationalization of educational policy and practice, and also the failure of comprehending educational value through educational honors, consumerism, and understanding the vision of educational promise defined by consumer capitalism and global economic competition. This opens the crucial question about the role of public schooling if it is no longer to be reduced to restricted economic ends. The principle educational question becomes the role of education for the global public good. It also relates to the crises facing the humanities at all levels of institutionalized education as under the neoliberal regime, subjects less likely to result in immediate financial return are defunded and devalued.

This book has criticized the neoliberal assumptions fostered by VP, historicized this contemporary movement in educational philanthropy, and finally, suggested that these problems with neoliberal VP belie a much deeper crisis in educational thought and practice involving economism across the political spectrum. Both

Marxian economism and neoliberal economism ought to be rejected in favor of a revitalized thinking of educational obligation and the educational contract. The current trend toward instrumentalizing education, treating students and teachers as clients and consumers, commodifying knowledge, as well as the various forms of privatizing, union-busting, and "turning around" schools all based in business language and assumptions ought to be halted. As well, it is time to end the anti-democratic concentrated power of privatized educational governance and policy by VPs by nationalizing or even globalizing the existing vast wealth of philanthropic foundations and restructuring the tax laws to stop publicly subsidizing the undermining of public governance. Literature on general economy including Mauss can offer valuable insights in the effort to depart from economism, revalue teaching and learning, rethink educational obligations, and remake educational institutions as part of an effort to expand global justice and public democracy.

# CONCLUSION

The chapters in this book have illustrated how educational philanthropy has been recently remade on the model of venture capital, and how this is part of the broader neoliberal remaking of public education that most significantly advances a privatization and deregulation agenda. Venture philanthropists in education have been pushing vouchers, charter schools, scholarship tax credits (neovouchers) and funding the infrastructure of the school privatization movement from think tanks and lobbying groups to political organizations to scholarship and publicity to grassroots "Astroturf" campaigns. As well, the venture philanthropists are behind the expansion of standardized test-based measures of educational value, attempts to transform educational leadership and teacher education in anti-intellectual and anti-critical formats, and the modeling of reform on corporate culture and ideals.

I have suggested that the shift from "scientific" to VP ought to be understood not only as part of the expansion of neoliberal ideology into education, the importation of the venture capital model into different domains, and the diminishing sense of the public but also in terms of the transformations from a production-oriented industrial economy to a consumption-oriented service economy. As the U.S. economy has become increasingly "frivolous"—that is, dependent upon the ever greater expansion of manufactured consumer desires rather than the use-value of commodities, "frivolous" giving in the form of charity without strings has appeared in need of eradication by VP. Instead, VP seeks to rationalize all aspects of giving and the control of the giving process. I call into question the economic rationale for public schooling based on the unlimited growth of a consumer economy. Also, I suggested that neoliberal educational reform should be understood as symptomatic of a more fundamental tendency toward economism across the political spectrum and that a revaluing of educational obligation offers a hopeful public alternative to the anti-public tendencies of thinking education through restricted economic exchange. This conclusion builds on these chapters by drawing the lens back yet farther to ask what is promised by VP and neoliberal educational reform more generally.

Neoliberal educational reform promises the student the dream-world of consumer luxury found in consumer society—a promise dominating educational policy. The call and implementation of high stakes standardized testing, the standardization of curriculum, the vocationalization of schooling, hyper-rationalization of all aspects of schooling, the calls to treat public schooling like a market—all of this is justified as leading to the promise of greater and greater consumption. The promise is based on educational exchange. The individual who works hard learning that, which others have determined is important, is ultimately promised the exchange of grades for higher schooling and then promised the exchange of grades and graduation for work, promised the exchange of work for greater income, and ultimately, promised the ability to maximize the self through the consumption of ever more goods and services. Within this view, knowledge is treated like cash to be earned and then exchanged for educational honors and eventually economic rewards. More recently, this logic has expanded and become even more explicit through both teacher bonus pay and payment to students for grades as well as the linkage of teacher and administrator preparation to student test scores. In these examples, testing is used as if it is a neutral and objective "market," serving as the medium for the acquisition of, display, and exchange, rewarding talent and work. Of course, not only are the politics of knowledge behind the framing, selection, and organization of knowledge denied in this view, but the broader public implications of learning and the public role of schools is evacuated in favor of a metaphor of markets and consumption.

This explicit equation of educational activity with the promise of greater and greater levels of consumption is made to the nation as well as to the individual. The nation's school kids must work hard in schools to compete globally with other nations to maintain or increase our nation's consuming capacity. In this zero-sum game, the losing poorer nations end up with the role of doing the brutal labor necessary to produce the commodities in the retail stores of the richer nations. This educational promise has no way of dealing with the global race to the bottom for cheaper and cheaper labor and worsened labor conditions, nor does it deal with the structural push for cheaper and cheaper labor domestically. As some, such as

Stanley Aronowitz argues, expanded higher education enrollment becomes a way to disguise unemployment. As the U.S. economy has entered crisis with high unemployment, we find the largest VP retooling their agenda to focus specifically on higher education preparation. In line with this, the Obama administration has on the one hand taken the positive step of expanding student loans and reforming the exploitative system of privatized high interest student loan servicing (though in reality, a far better goal would be to work for universal free higher education as exists in other industrialized nations). Yet, on the other hand, the administration has introduced the "race to the top" program with its use of much needed federal funds to push privatization in the form of charters. This program follows the logic of NCLB with the regressive use of federal money to reward only those who get in line with the agenda. While NCLB distributes resources to those with the capital and cultural capital to score well on tests, "race to the top" rewards those willing to loosen restrictions on charter school expansion. This amounts to an assault on teachers unions and local school boards. What should not be missed is that the unfortunate language used for this presumes an exclusionary national competition standing starkly at odds with values of egalitarianism. But it also references the standard left accusation about the global economy that neoliberal capitalism fosters a "race to the bottom" as deregulation of controls over capital and privatization force countries to compete for low paying, no benefit jobs. The name of the program is both a denial of the reality of the global "race to the bottom" and an admission that the highest aspiration of these market based school reforms is the pathetic inclusion of students in this losing game. School reform for a progressively minded administration would at the very least put forward educational goals framed through the language of collective social aspirations. That is, educational projects could be initiated to make public forms of community and global improvement the basis for learning rather than continuing the Bush emphasis on individualized academic achievement narrowly defined. Such projects could give students the language, theoretical and interpretive skills to comprehend the effects of oppressive social forces that they experience day-to-day and the intellectual tools to work to transform them.

In both the case of the individual and the nation, the market-based promise of the continuing neoliberal reform is driven by the idealized form of consumption, that is, luxury—luxury understood through greater and more glorious forms of commodity acquisition and consumer activity. The social cost of consumerism as the core educational promise includes not merely a crisis of meaning, the alienation of the individual from the self, from nature, and from others, but it also empties out the political and ethical possibilities of education as the only vision of social improvement becomes the individual promise of consumer commodity acquisition. The social cost is even greater in the sense that constantly increasing economic growth as the guiding force of the economy guarantees ecological devastation and vast inequalities in standards of living. Unlimited economic growth as the economic and educational goal is an utterly unsustainable ideal which will only result in the current heading toward ecological collapse, natural disaster, and human catastrophe. In the neoliberal view, there is no alternative to the present, and so the aim of education is to enforce the existing order—an order which is misrepresented as natural and inevitable rather than as being currently enacted through policy. There is a kind of double violence involved in the neoliberal educational project. On the one hand, it naturalizes the neoliberal economic uses of education thereby naturalizing and misrepresenting as inevitable what are, in fact, human policies and priorities for profit accumulation above all else. On the other hand, neoliberal educational reform undermines the development of critical forms of education that would serve as the conditions for a public with the intellectual tools and dispositions to collectively solve public problems and move the society on a different heading from the dead-end guarantee of unlimited growth. The anti-critical and anti-intellectual forms of schooling that are being fostered by neoliberal educational reform deprive citizens of the capacities to imagine and enact alternatives for the future in part because they fail to teach how to criticize and analyze the assumptions and ideologies undergirding claims to truth. The dispositions of curiosity, disciplined creativity, and investigation under attack by neoliberal educational reforms are crucial for fostering in citizens the skills of interpretation and social intervention and for imagining and also enacting alternatives to the present.

## WHAT IS TO BE DONE?

Venture philanthropy needs to be understood as part of an onslaught of neoliberal educational reforms the assumptions of which ought to be seen as utterly discredited. The idealization of markets and deregulation and privatization appear as untenable fundamental ideals as the public sector steps in to save private industry from itself, rescuing industry after industry. Venture philanthropy ought to be seen as another failed business-led reform of public schooling and should be considered in relation to the long history of business-led school reform that has largely resulted in the current educational inequalities. Public schooling needs to be revalued as a public institution with public purposes that are primarily for the fostering of public intellectual activity linked to social reconstruction and the activities of engaged public life.

The promise of educational exchange that is premised on the promise of consumer luxury is not only based in ecologically, politically, and economically destructive assumptions. It also fails to grasp what I call in Chapter 6 the originary pledge in educational exchange. As I contend, the students who pledge their lives and time to the public school system are owed material and symbolic counter pledges by the whole society. Neoliberal education and VP presume that the educational exchange begins with the student owing the society for educational and economic opportunities and resources whether or not the society has historically and presently ponied up. Consequently, these approaches to school improvement begin by assuming that the basic problems stem not from the historical and systematized material and symbolic violence of unequal educational exchange but rather from a failure of individual discipline on the parts of the students and teachers; and they assume that the primary task is one of enforcing the official knowledge and program through various coercive measures. Instead, educational improvement ought to begin with the assumption that the school and the society owe the student an enormous debt incurred by the life and presence of the student. The counter pledge, the educational contract from the school and society can only be fulfilled materially by accounting for the history of unequal material exchange not only in terms of educational resources but in terms of material exchange generally. This means that schools should prepare students to engage material reality not only through the tools of mathematics and science, literacy, and social analysis. This means that part of what public schools should do as public democratic institutions is prepare students to play an active role in democratizing material relations outside of schools so that future citizens are equipped to participate in economic decisions about production and consumption that affect all citizens materially. As well, the educational contract from the school and the society can only be fulfilled symbolically by offering students the tools to analyze and interpret cultural, ideological, and linguistic formations in relation to broader economic and political forces. Public schools are obligated to prepare citizens in the making for practices of interpretation and meaning-making as public individuals whose actions and signifying practices have public import.

Educators and cultural workers as well as student, policy makers, union organizers and those committed to education for social justice need to work for the following:

1. To end tax breaks for foundations and erect a wall between giving and the use of money for education as part of a larger movement against business-driven educational reform. If money is given by private interests, then public control ought to be fully retained over the use of educational spending.

2. To stop the application of economism to educational reform. With the collapse of neoliberal assumptions, we should stop applying business metaphors and logic to educational thinking derived from discredited market fundamentalism. Metaphors not only of deregulation and privatization but also market-based framings of "competition and choice," "monopoly," "turnaround," and "efficiency" need to be dropped in favor of public language and assumptions including equality, the public interest, public pedagogy, and a renewed language of educational obligation grounded in public democratic values. I have suggested in this book that we might develop a notion of the educational contract that reworks the neoliberal view of educational obligation. The "measure" of educational progress ought not to be test scores but rather social progress measured by the dismantling of oppressive institutions and practices and the making of institutions and practices that provide contexts for egalitarian social relations, democratic debate and dialogue in strong public spheres.

3. We should nationalize foundation wealth and give it to public educational authorities. Consider the case of the Bill and Melinda Gates Foundation wealth. Microsoft's private wealth was the result first of billions of dollars of public subsidy for the development of the computer industries, the internet, and information technology. This has been a case of socialized funding and privatized profits. Once the public paid to develop the industries, the profit from the technologies was handed over to private companies. The public then subsidized super-rich individuals giving tax breaks for foundations that essentially subsidized these private individuals to take control over educational policy steering. In essence, the public has paid to give control over public policy to elites who were already the beneficiaries of public funding for high-tech development, real estate riches, and retail fortunes. This

circuit of privatization must be ended, and public control over public policy formation must be democratized.

4. We should understand that Gates and the other venture philanthropists are neither generous nor disinterested and then take a cue from Pierre Bourdieu:

> The purely speculative and typically scholastic question of whether generosity and disinterestedness are possible should give way to the political question of the means that have to be implemented in order to create universes in which, as in gift economies, people have an interest in disinterestedness and generosity, or rather, are durably disposed to respect these universally respected forms of respect for the universal.[1]

# Coda: Obama's Betrayal of Public Education? Arne Duncan and the Corporate Model of Schooling

*Henry A. Giroux and Kenneth J. Saltman*

[www.truthout.org, December 17, 2008]

Since the 1980s, but particularly under the Bush administration, certain elements of the religious right, the corporate culture, and the Republican right wing have argued that free public education represents either a massive fraud or a contemptuous failure. Far from a genuine call for reform, these attacks largely stem from an attempt to transform schools from a public investment to a private good, answerable not to the demands and values of a democratic society but to the imperatives of the marketplace. As the educational historian David Labaree rightly argues, public schools have been under attack in the last decade "not just because they are deemed ineffective but because they are public."[1] Right-wing efforts to disinvest in public schools as critical sites of teaching and learning and govern them according to corporate interests is obvious in the emphasis on standardized testing, the use of top-down curricular mandates, the influx of advertising in schools, the use of profit motives to "encourage" student performance, the attack on teacher unions and modes of pedagogy that stress rote learning and memorization. For the Bush administration, testing has become the ultimate accountability measure, belying the complex mechanisms of teaching and learning. The hidden curriculum is : that testing be used as a ploy to de-skill teachers by reducing them to mere technicians; that students be similarly reduced to customers in the marketplace rather than as engaged, critical learners and that always underfunded public schools fail so that they can eventually be privatized. But, there is an even darker side to the reforms initiated under the Bush administration and now used in a number of school systems throughout the country. As the logic of the market and "the crime complex"[2] frame the field of social relations in schools, students are subjected to three particularly offensive policies, defended by school authorities and politicians under the rubric of school safety. First, students are increasingly subjected to zero-tolerance policies that are used primarily to punish, repress, and exclude them. Second, they are increasingly absorbed into a "crime complex" in which security staff, using harsh disciplinary practices, now displace the normative functions teachers once provided both in and outside of the classroom.[3] Third, more and more schools are breaking down the space between education and juvenile delinquency, substituting penal pedagogies for critical learning and replacing a school culture that fosters a discourse of possibility with a culture of fear and social control. Consequently, many youth of color in urban school systems, because of harsh zero-tolerance polices, are not just being suspended or expelled from school. They are being ushered into the dark precincts of juvenile detention centers, adult courts, and prison. Surely, the dismantling of this corporatized and militarized model of schooling should be a top priority under the Obama administration. Unfortunately, Obama has appointed as his secretary of education someone who actually embodies this utterly punitive, anti-intellectual, corporatized and test-driven model of schooling.

Barack Obama's selection of Arne Duncan for secretary of education does not bode well either for the political direction of his administration nor for the future of public education. Obama's call for change falls flat with this appointment, not only because Duncan largely defines schools within a market-based and penal model of pedagogy, but also because he does not have the slightest understanding of schools as something other than adjuncts of the corporation at best or the prison at worse. The first casualty in this scenario is a language of social and political responsibility capable of defending those vital institutions that expand the rights, public goods and services central to a meaningful democracy. This is especially true with respect to the issue of public schooling and the ensuing debate over the purpose of education, the role of teachers as critical intellectuals, the politics of the curriculum, and the centrality of pedagogy as a moral and political practice.

Duncan, CEO of the Chicago Public Schools, presided over the implementation and expansion of an agenda that militarized and corporatized the third largest school system in the nation, one that is about 90 percent poor and nonwhite. Under Duncan, Chicago took the lead in creating public schools run as military academies, vastly expanded draconian student expulsions, instituted sweeping surveillance practices, advocated a growing police presence in the schools, arbitrarily shut down entire schools, and fired entire school staffs. A recent report, "Education on Lockdown," claimed that partly under Duncan's leadership "Chicago Public Schools (CPS) has become infamous for its harsh zero tolerance policies. Although there is no verified positive impact on safety, these policies have resulted in tens of thousands of student suspensions and an exorbitant number of expulsions."[4] Duncan's neoliberal ideology is on full display in the various connections he has established with the ruling political and business elite in Chicago.[5] He led the Renaissance 2010 plan, which was created for Mayor Daley by the Commercial Club of Chicago—an organization representing the largest businesses in the city. The purpose of Renaissance 2010 was to increase the number of high quality schools that would be subject to new standards of accountability—a code word for legitimating more charter schools and high stakes testing in the guise of hard-nosed empiricism. Chicago's 2010 plan targets 15 percent of the city district's alleged underachieving schools in order to dismantle them and open 100 new experimental schools in areas slated for gentrification. Most of the new experimental schools have eliminated the teacher union. The Commercial Club hired corporate consulting firm A.T. Kearney to write Ren2010, which called for the closing of 100 public schools and the reopening of privatized charter schools, contract schools (more charters to circumvent state limits), and "performance" schools. Kearney's web site is unapologetic about its business-oriented notion of leadership, and one that John Dewey thought should be avoided at all costs. It states, "Drawing on our program-management skills and our knowledge of best practices used across industries, we provided a private-sector perspective on how to address many of the complex issues that challenge other large urban education transformations."[6]

Duncan's advocacy of the Renaissance 2010 plan alone should have immediately disqualified him for the Obama appointment. At the heart of this plan is a privatization scheme for creating a "market" in public education by urging public schools to compete against each other for scarce resources and by introducing "choice" initiatives so that parents and students will think of themselves as private consumers of educational services.[7] As a result of his support of the plan, Duncan came under attack by community organizations, parents, education scholars and students. These diverse critics have denounced it as a scheme less designed to improve the quality of schooling than as a plan for privatization, union-busting, and the dismantling of democratically elected local school councils. They also describe it as part of neighborhood gentrification schemes involving the privatization of public housing projects through mixed finance developments.[8] (Tony Rezko, an Obama and Blagojevich campaign supporter, made a fortune from these developments along with many corporate investors.) Some of the dimensions of public school privatization involve Renaissance schools being run by subcontracted for-profit companies—a shift in school governance from teachers and elected community councils to appointed administrators coming disproportionately from the ranks of business. It also establishes corporate control over the selection and model of new schools, giving the business elite and their foundations increasing influence over educational policy. No wonder that Duncan had the support of David Brooks, the conservative op-ed writer for The New York Times.

One particularly egregious example of Duncan's vision of education can be seen in the conference he organized with the Renaissance Schools Fund. In May 2008, the Renaissance Schools Fund, the financial wing of the Renaissance 2010 plan operating under the auspices of the Commercial Club, held a symposium, "Free to Choose, Free to Succeed: The New Market in Public Education," at the exclusive private club atop the Aon Center. The event was held largely by and for the business sector, school privatization advocates, and others already involved in Renaissance 2010, such as corporate foundations and conservative think tanks. Significantly, no education scholars were invited to participate in the proceedings, although it was heavily attended by fellows from the pro-privatization Fordham Foundation and featured speakers from various school choice organizations and the leadership of corporations. Speakers clearly assumed the audience shared their views.

Without irony, Arne Duncan characterized the goal of Renaissance 2010 creating the new market in public education as a "movement for social justice." He invoked corporate investment terms to describe reforms explaining that the 100 new schools would leverage influence on the other 500 schools in Chicago. Redefining schools as stock investments he said, "I am not a manager of 600 schools. I'm a portfolio manager of 600 schools and I'm trying to improve the portfolio." He claimed that education can end poverty. He explained that having a sense of altruism is important, but that creating good workers is a prime goal of educational reform and that the business sector has to embrace public education. "We're trying to blur the lines between the public and the private," he said. He argued that a primary goal of educational reform is to get the private sector to play a huge role in school change in terms of both money and intellectual capital. He also attacked the Chicago Teachers Union (CTU), positioning it as an obstacle to business-led reform. He also insisted that the CTU opposes charter schools (and, hence, change itself), despite the fact that the CTU runs ten such schools under Renaissance 2010. Despite the representation in the popular press of Duncan as conciliatory to the unions, his statements and those of others at the symposium belied a deep hostility to teachers unions and a desire to end them (all of the charters created under Ren2010 are deunionized). Thus, in Duncan's attempts to close and transform low-performing schools, he not only reinvents them as entrepreneurial schools, but, in many cases, frees "them from union contracts and some state regulations."[9] Duncan effusively praised one speaker, Michael Milkie, the founder of the Nobel Street charter schools, who openly called for the closing and reopening of every school in the district precisely to get rid of the unions. What became clear is that Duncan views Renaissance 2010 as a national blueprint for educational reform, but what is at stake in this vision is the end of schooling as a public good and a return to the discredited and tired neoliberal model of reform that conservatives love to embrace.

In spite of the corporate rhetoric of accountability, efficiency and excellence, there is to date no evidence that the radical reforms under Duncan's tenure as the "CEO" of Chicago Public Schools have created any significant improvement. In part, this is because the Chicago Public Schools and the Renaissance Schools Fund report data in obscurantist ways to make traditional comparisons difficult if not impossible.[10] And, in part, examples of educational claims to school improvement are being made about schools embedded in communities that suffered dislocation and removal through coordinated housing privatization and gentrification policies. For example, the city has decimated public housing in coveted real estate enclaves, dispossessing thousands of residents of their communities. Once the poor are removed, the urban cleansing provides an opportunity for Duncan to open a number of Renaissance Schools, catering to those socioeconomically empowered families whose children would surely improve the city's overall test scores. What are alleged to be school improvements under Ren2010, rest on an increase in the city's overall test scores and other performance measures that parodies the financial shell game corporations used to inflate profit margins—and prospects for future catastrophes are as inevitable. In the end, all Duncan leaves us with is a Renaissance 2010 model of education that is celebrated as a business designed "to save kids" from a failed public system. In fact, it condemns public schooling, administrators, teachers and students to a now outmoded and discredited economic model of reform that can only imagine education as a business, teachers as entrepreneurs, and students as customers.[11]

It is difficult to understand how Barack Obama can reconcile his vision of change with Duncan's history of supporting a corporate vision for school reform and a penchant for extreme zero-tolerance polices—both of which are much closer to the retrograde policies hatched in conservative think tanks such as the Heritage Foundation, Cato Institution, Fordham Foundation, American Enterprise Institute, than to the values of the many millions who voted for the democratic change he promised. As is well known, these think tanks share an agenda not for strengthening public schooling, but for dismantling it and replacing it with a private market in consumable educational services. At the heart of Duncan's vision of school reform is a corporatized model of education that cancels out the democratic impulses and practices of civil society by either devaluing or absorbing them within the logic of the market or the prison. No longer a space for relating schools to the obligations of public life and social responsibility to the demands of critical and engaged citizenship, schools in this dystopian vision legitimate an all-encompassing horizon for producing market identities, values, and those privatizing and penal pedagogies that both inflate the importance of individualized competition and punish those who do not fit into its logic of pedagogical Darwinism.[12]

In spite of what Duncan argues, the greatest threat to our children does not come from lowered standards, the absence of privatized choice schemes, or the lack of rigid testing measures that offer the aura of accountability. On the contrary, it comes from a society that refuses to view children as a social investment, consigns 13 million children to live in poverty, reduces critical learning to massive testing programs, promotes policies that eliminate most crucial health and public services, and defines rugged individualism through the degrading celebration of a gun culture, extreme sports, and the spectacles of violence that permeate corporate-controlled media industries. Students are not at risk because of the absence of market incentives in the schools. Young people are under siege in American schools because, in the absence of funding, equal opportunity, and real accountability, far too many of them have increasingly become institutional breeding grounds for racism, right-wing paramilitary cultures, social intolerance, and sexism.[13] We live in a society in which a culture of testing, punishment, and intolerance has replaced a culture of social responsibility and compassion. Within such a climate of harsh discipline and disdain for critical teaching and learning, it is easier to subject young people to a culture of faux accountability or put them in jail rather than to provide the education, services, and care they need to face problems of a complex and demanding society.[14] What Duncan and other neoliberal economic advocates refuse to address is what it would mean for a viable educational policy to provide reasonable support services for all students and viable alternatives for the troubled ones. The notion that children should be viewed as a crucial social resource— one that represents, for any healthy society, important ethical and political considerations about the quality of public life, the allocation of social provisions and the role of the state as a guardian of public interests—appears to be lost in a society that refuses to invest in its youth as part of a broader commitment to a fully realized democracy. As the social order becomes more privatized and militarized, we increasingly face the problem of losing a generation of young people to a system of increasing intolerance, repression, and moral indifference. It is difficult to understand why Obama would appoint as secretary of education someone who believes in a market-driven model that has not only failed young people, but given the current financial crisis has been thoroughly discredited. Unless Duncan is willing to reinvent himself, the national agenda he will develop for education embodies and exacerbates these problems and, as such, it will leave a lot more kids behind than it helps.

# Appendix

## Chapter 1

This appendix is abridged. For a more complete appendix of tables go to www.book-smarts.net

**Tables 1.1.A–D**  Top Ten Foundations Awarding Grants for Education (circa 2007)

A. Overall Giving

| Foundation Name | State | Total Dollars Awarded | Grants Awarded |
|---|---|---|---|
| Bill & Melinda Gates Foundation | WA | $309,984,857 | 155 |
| The Andrew W. Mellon Foundation | NY | $201,470,275 | 347 |
| The William and Flora Hewlett Foundation | CA | $172,921,447 | 99 |
| Walton Family Foundation, Inc. | AR | $119,169,250 | 236 |
| Lilly Endowment, Inc. | IN | $116,813,807 | 21 |
| The Duke Endowment | NC | $94,412,935 | 29 |
| The Community Foundation | DC | $65,594,329 | 366 |
| W.K. Kellogg Foundation | MI | $60,303,697 | 51 |
| Carnegie Corporation of New York | NY | $49,451,588 | 80 |
| The Starr Foundation | NY | $43,986,830 | 124 |

*Source*: http://foundationcenter.org/findfunders/statistics/pdf/04_fund_sub/2007/50_found_sub/f_sub_b_07.pdf

## CHAPTER 3

**Tables 3.1.A–F** Educational Leadership Training Projects (Broad Foundation)

A. Teachers

| | 2007 | 2006 | 2005 | 2004 | 2003 | 2002 | 2001 |
|---|---|---|---|---|---|---|---|
| National Institute for Excellence in Teaching | $825,878 | $381,263 | | | | | |
| Teach for America | $2,779,000 | $2,523,150 | $624,630 | $523,350 | $501,000 | $394,360 | $135,000 |
| TOTAL | $3,604,878 | $2,904,413 | $624,630 | $523,350 | $501,000 | $394,360 | $135,000 |

B. Principals

| | 2007 | 2006 | 2005 | 2004 | 2003 | 2002 | 2001 |
|---|---|---|---|---|---|---|---|
| Boston Plan for Excellence | $250,000 | | | | $776,271 | | |
| Boston Public Schools | | $650,000 | | | | | |
| Fund for Public Schools | $147,288 | $625,000 | $875,000 | | $1,000,000 | | |
| Gwinnett County Public Schools | $624,108 | | | | | | |
| Long Beach Unified School District | $157,750 | | | | | | |
| National Center on Education and the Economy | | | $600,000 | $600,000 | | $800,000 | $800,000 |
| New Leaders for New Schools | $375,000 | $375,000 | $720,000 | $2,000,000 | $2,250,000 | $1,218,000 | $1,056,000 |
| New Teacher Project | | | $120,000 | $150,000 | | | |
| San Diego City Schools | | | | $150,000 | $150,000 | | |
| School District of Philadelphia | $549,217 | $550,000 | $1,050,000 | | | | |
| TOTAL | $549,217 | $550,000 | $1,050,000 | $150,000 | $150,000 | $2,018,000 | $1,856,000 |

C. Superintendents

| | 2007 | 2006 | 2005 | 2004 | 2003 | 2002 | 2001 |
|---|---|---|---|---|---|---|---|
| Aspen Institute | $10,000 | | | | | | |
| Aurora Schools | | | $60,000 | $60,000 | $19,438 | $40,562 | |
| Benton Harbor Public Schools | | $10,000 | | | | | |
| Broad Center for Urban Superintendents | | | $1,040,620 | $2,039,791 | $879,935 | $1,063,755 | $502,015 |
| Christina Partners for Excellence | $10,000 | | | | | | |
| Council for the Great City Schools | | | | | $80,000 | | |
| Fort Wayne Community Schools | $5,000 | $10,000 | | | | | |
| Houston Independent School District | | | | $75,131 | $152,094 | | |
| Los Angeles Unified School District | | | | | | $100,000 | |
| Montgomery Public Schools | | $10,000 | | | | | |
| Paterson City Schools | $5,000 | $10,000 | | | | | |
| TOTAL | $5,000 | $20,000 | $1,040,620 | $75,131 | $152,094 | $100,000 | $502,015 |

D. School Boards

| | 2007 | 2006 | 2005 | 2004 | 2003 | 2002 | 2001 |
|---|---|---|---|---|---|---|---|
| American Productivity and Quality Center | | | | | | | $590,030 |
| Center for Reform of School Systems | $1,633,819 | $1,319,635 | $1,209,189 | $564,043 | $501,166 | $471,256 | $113,420 |
| Children's Defense Fund | | | | | | | $24,000 |
| Clark County School District | | $50,000 | | | | | |
| Greater Atlanta Chamber Foundation | | | | | | | $20,000 |
| National School Boards Association | | | $25,000 | | $25,000 | | |
| Oakland Unified School District | | | | | | $6,000 | |
| TOTAL | $1,633,819 | $50,000 | $25,000 | $564,043 | $25,000 | $6,000 | $20,000 |

**Tables 5.1.A–H** System Redesign Projects and Initiatives

A. General System Redesign

| | 2007 | 2006 | 2005 | 2004 | 2003 | 2002 | 2001 |
|---|---|---|---|---|---|---|---|
| Children's First Fund | $600,000 | | | | | | |
| Christina School District | | | $500,000 | $13,000 | $39,000 | | |
| Council of the Great City Schools | $82,204 | | $65,174 | $139,346 | | | |
| Fund for Public Schools | $2,731,045 | $1,800,000 | $1,838,204 | | | | |
| New Visions for New Schools | | | | | $1,640,000 | $600,200 | |
| New York City Department of Education | | $750,000 | $164,420 | | | | |
| Pasadena Unified School District | | | | | | | $8,650 |
| Providence Public Schools | | | | | $29,150 | | |
| TOTAL | $3,413,249 | $2,550,000 | $2,567,798 | $152,346 | $1,708,150 | $606,571 | $8,650 |

# NOTES

## INTRODUCTION

1. Slavoj Zizek, *Violence* New York: Picador 2008, pp. 20–22.
2. See Robert Arnove, *Philanthropy and Cultural Imperialism* and William Watkins, *The White Architects of Black Education*.
3. Neoliberalism involves redistributing public goods to private controls while espousing market triumphalism. As David Harvey explains it is a project of class warfare. Chapter 1 goes into greater depth about both neoliberalism and neoliberal education reforms as expressed through VP.
4. Notable early exceptions to this include Frederick Hess' edited collection *With the Best of Intentions* from a neoliberal perspective and the liberal work of Janelle Scott such as "The Politics of Venture Philanthropy in Charter School Policy and Advocacy" *Educational Policy* 23(1), 106–136 and the work of Rick Cohen of the Center for Responsive Philanthropy discussed in Chapter 1. Mike and Susan Klonsky's book *Small Schools.* New York: Routledge (2008) and Philip Kovacs's scholarship on Gates stand out as some of the rare critical work on VP. See for example, Philip Kovacs and H.K. Christie "The Gates Foundation and the Future of U.S. Public Education: A Call for Scholars to Counter Misinformation Campaigns" *The Journal for Critical Education Policy Studies* 6(2), December 2008.
5. See Henry Giroux's chapter in *Disturbing Pleasures* New York: Routledge 1992 for a brilliant criticism of the way Benetton conflated political agency with consumerism.

## CHAPTER 1

1. I refer to the traditions of critical pedagogy and critical theory as well as to the varieties of thought characterized by radical democracy with its focus on the expansion of egalitarian social relations and its emphasis on the priority of culture as well as the redistributive economic theories of the socialist tradition.
2. In this chapter, I will look broadly at the varieties of right-wing approaches to corporatization making a division between fiscal and cultural conservatives. For a more elaborate discussion of the varieties of rightist thought including neoliberalism, neoconservatism, religious fundamentalism and authoritarian populism see, Michael Apple, *Educating the Right Way* (2nd ed.) New York: Routledge, 2006.
3. Despite the centrality of broader social struggle and structural transformation, critical perspectives on corporatization are hardly identical and tend to map to

the different political, economic, and cultural referents for theorizing the phenomenon. Criticism of corporatization can be found grounded in perspectives including Neo-marxist (Apple and Lipman), Marxist (McLaren, Hill, etc.), critical theory, radical democracy (Giroux, Aronowitz, and Trend), Foucauldian (Ball), poststructuralist, pragmatist (Molnar and Boyles), and anarchist (Spring and Gabbard) with many authors drawing on multiple traditions. There is not space here to elaborate fully these different political referents for the criticism of corporatization. However, some of the major disagreements involve questions of the centrality of class and economics as opposed to the centrality of culture, identity, difference, and language, as well as debates over the possibilities for the role of the state in public versus private spheres. Many of the debates that face criticalists often have a legacy that recapitulates many of the central debates initiated by Marx and carried down from the Marxian tradition, though the minority of criticalists define themselves as Marxists. Other disagreements in the perspective on corporatization are a matter of methodological emphasis on institutional analysis as opposed to broader structural analysis. Joel Spring—writing from the anarchist tradition—and Alex Molnar—grounded in Deweyan progressivism—each tend to offer valuable institutionally focused analyses. These are highly compatible with other perspectives that emphasize the situating of corporatization in relation to broader structural forces and trends. All critics of corporatization from a critical perspective view the corporatization of public schooling as an expression of neoliberal capitalism.

4. This critical view of hegemonic struggle can be traced from Gramsci to Althusser and can be seen in the critical response to corporatization in contemporary writers such as Apple, Giroux, Lipman, Saltman, Leistyna, and numerous others. See Gramsci's *The Prison Notebooks*; Althusser's "Ideology and Ideological State Apparatuses: Notes Towards an Investigation."

5. Joel Bakan's *The Corporation*, New York: Free Press, 2004, p. 57.

6. See Farhad Manjoo, "Microsoft's Media Monopoly" Salon.com, October 29, 2002 available at http://dir.salon.com/story/tech/feature/2002/10/29/microsoft_media_one/index.html.

7. See Joel Spring's excellent history in *Educating the Consumer-Citizen*.

8. See David Harvey's *A Brief History of Neoliberalism*. In education, see Kenneth Saltman's *Collateral Damage* and *Strange Love*, Michael Apple's *Educating the Right Way*, and Henry Giroux's *The Terror of Neoliberalism*.

9. See Alex Molnar, Gary Miron, Jessica Urschel, "Profiles of For-Profit Educational Management Organizations: Eleventh Annual Report" September 2009 Commercialism in Education Research Unit available at http://epicpolicy.org/files/08–09%20profiles%20report.pdf.

10. See Kenneth J. Saltman's *Capitalizing on Disater*, Boulder: Paradigm Publishers and Kenneth J. Saltman's *Schooling and the Politics of Disaster*, New York: Routledge.

11. For the clearest and most up-to-date coverage of the terrain and scope of public school privatization and commercialization, see Alex Molnar's annual reports on school commercialism online at the Educational Policy Studies Laboratory available at www.schoolcommercialism.org as well as Alex Molnar's *School Commercialism*, New York: Routledge, 2005. Molnar's

theoretical perspective draws on Deweyan pragmatism and the institutional analysis typified by Edward Herman and Noam Chomsky's media analysis that focuses on the active role of the public relations industry in "manufacturing consent" to business dominance. Joel Spring's important historically sweeping *Educating the Consumer-Citizen,* Lawrence Earlbaum and Associates, 2003 likewise offers essential institutional analysis. Such historical institutional analyses are crucial yet incomplete in that they need to be coupled with a more comprehensive cultural theorization as well as broader situating within global economic structural dynamics. For important recent scholarships on a range of issues involved in privatization, see, for example, Deron Boyles (editor) in *Schools or Markets?: Commercialism, Privatization and School-Business Partnerships,* New York: Lawrence Earlbaum and Associates, 2004; Joel Spring's *Educating the Consumer-Citizen,* Lawrence Earlbaum and Associates, 2003; and Alfie Kohn and Patrick Shannon's *Education, Inc.,* Heinemann, 2002. See also Kenneth Saltman, "Essay Review of Education, Inc." Teachers College Record, 2003.

12. See Molnar, *School Commercialism.*
13. State run schools in capitalist nations being a "site and stake" of struggle for hegemony appears in the work of Antonio Gramsci and is developed from Gramsci by Louis Althusser in, for example, his "Ideology and Ideological State Apparatuses" essay in *Lenin and Philosophy and Other Essays,* New York: Monthly Review Press, 2001 and more explicitly with reference to Gramsci in *Machiavelli and Us,* New York: Verso, 1997. The limitations of the reproduction theories have been taken up extensively and importantly, for example, with regard to the theoretical problems of Marxism including the legacies of scientism, class reductionism, economism, and so on. See, for example, Stanley Aronowitz and Henry Giroux *Education Still Under Siege,* Westport: Bergin and Garvey, 1989. Despite these limitations, Althusser's work appears important for theorizing the state at the present juncture.
14. See, for example, Giroux's important discussion of the politics of "No Child Left Behind" in *Abandoned Generation,* New York: Palgrave, 2003.
15. Neoliberalism in education has been taken up extensively by a number of authors. A very partial and incomplete and U.S.-focused list includes the following: Kenneth J. Saltman, *Collateral Damage*; Robin Goodman and Kenneth J. Saltman, *Strange Love, Or How We Learn to Stop Worrying and Love the Market*; Kenneth J. Saltman and David Gabbard (eds.) *Education as Enforcement: the Militarization and Corporatization of Schools*; Michael Apple, *Educating the Right Way*; Henry Giroux, *Abandoned Generation*; Henry Giroux, *The Terror of Neoliberalism*; David Gabbard and E. Wayne Ross, *Defending Public Schools.*
16. An excellent mapping of these conservatisms and others can be found in Michael Apple's *Educating the Right Way,* New York: Routledge, 2001.
17. David Harvey, *A Brief History of Neoliberalism,* Oxford: Oxford University Press, 2005, pp. 66–67.
18. John Chubb and Terry Moe, *Politics, Markets, and America's Schools.*
19. Jeffrey Henig, *Rethinking School Choice,* Princeton: Princeton University Press, 1994.
20. Stephen J. Ball, *Education plc,* New York: Routledge, 2007.

21. I take up these issues in greater detail in Kenneth J. Saltman, *The Edison Schools,* New York: Routledge 2005.

22. Within the field of education, the contemporary traditions of critical pedagogy and critical literacy continue to pursue and develop this.

## Chapter 2

1. The most significant collection of essays from mostly advocates of neoliberal venture philanthropy is Frederick Hess (ed.) *With the Best of Intentions,* Cambridge, MA: Harvard Education Press (2005). However, a largely unified perspective can be found on the websites and event materials, public relations materials and reports of the leading venture philanthropists such as Gates, Broad, and Walton as well as more narrowly defined ones such as the NSVF, CSGF, and business-oriented organizations such as Chicago's Renaissance Schools Fund which is the financial arm of the Renaissance 2010 plan, which was devised for the mayor by the Commercial Club of Chicago.

2. Richard Lee Colvin in Frederick Hess (ed.) "Chapter 1: A New Generation of Philanthropists and Their Great Ambitions" in *With the Best of Intentions,* Harvard Education Press, 2005, p. 23; Frederick Hess, "Conclusion" p. 295.

3. Richard Lee Colvin in Frederick Hess (ed.) "Chapter 1: A New Generation of Philanthropists and Their Great Ambitions" *With the Best of Intentions,* Harvard Education Press, 2005, p. 21.

4. Colvin, p. 27.

5. These points are modified from those made by Rick Cohen, "Strategic Grantmaking: Foundations and the School Privatization Movement" National Committee for Responsive Philanthropy, November 2007 available online at http://www.ncrp.org/index.php?option=com_ixxocart&Itemid=4 1&p=product&id=4&parent=3.

6. See, for example, Kenneth Saltman *Collateral Damage,* Lanham, MD: Rowman & Littlefield, 2000; *The Edison Schools,* New York: Routledge 2005;*Capitalizing on Disaster,* Boulder: Paradigm Publishers, 2007.

7. As economic doctrine neoliberalism calls for the privatization of public goods and services, the deregulation of markets, foreign direct investment, and monetarism. Neoliberalism represents an ideology of market fundamentalism in which the inevitably bureaucratically encumbered state can do no good, and markets must be relied upon to do what the state has formerly done. Neoliberalism imagines the social world as privatized and suggests that economic rationality ought to be expanded to every last realm. In this view, the public sector disappears as the only legitimate collectivities can be markets while the individual is principally defined as an economic actor, that is, a worker or consumer. The state in this view ought to use its power to facilitate markets. Democracy becomes an administrative matter best left to markets rather than to public deliberation.

8. Pierre Bourdieu, *Firing Back: Against the Tyranny of the Market,* New York: New Press, 2003.

9. See Joel Spring, *Educating the Consumer-Citizen,* New York: Lawrence Earlbaum, 2003.

10. See Dorothy Shipps, *School Reform Corporate Style: Chicago 1880–2000*, Lawrence, KS: University Press of Kansas, 2006.
11. Peter Frumkin, "Inside Venture Philanthropy" *Society* May/June 2003, p. 4.
12. Rick Cohen, p. 18.
13. Rick Cohen, p. 5.
14. Rick Cohen, p. 55.
15. Arnove quoted in Erkki Berndtson, "Review Essay: Power of Foundations and the American Ideology," *Critical Sociology* 33(2007) 575–587.
16. This is particularly clear in the work of Stephen and Abigail Thernstrom's *No Excuses: Closing the Racial Gap in Learning*, New York: Simon and Shuster 2003, which the Renaissance Schools Fund, the funding arm of Renaissance 2010 makes recommended reading to prospective "entrepreneurs" interested in starting schools under Chicago's Renaissance 2010.
17. Stanley Aronowitz, *Against Schooling for An Education That Matters*, Boulder: Paradigm Publishers, 2008, p. 53.
18. See Robert Arnove, *Philanthropy and Cultural Imperialism: The Foundations at Home and Abroad*, Bloomington: Indiana University Press, 1982.
19. Joan Roelofs, "Foundations and Collaboration," *Critical Sociology* 33(2007) 479–504.
20. Stanley N. Katz, "Philanthropy's New Math" *The Chronicle of Higher Education*, 1F2 2007 supp., p. 3.
21. Steven Strauss "The Robber Baron as Lord Bountiful Bill Gates and the Capitalist Philanthropy Scam," *Freedom Socialist* 27(4) August–September 2006.
22. This point is made by multiple scholars of foundations including Joan Roelofs and Robert Arnove.
23. Paying teachers bonuses linked to student test performance exacerbates the ways that standardized testing contributes to the unequal distribution of cultural capital and the ways that class and culturally specific knowledge is treated as universally valuable. In addition to fostering "teaching to the test," this undermines the possibility of critical forms of teaching and learning that connect claims to truth to questions of power and authority and broader structures of power.
24. Rick Cohen, III.
25. Cohen, V.
26. Cohen, p. 12.
27. Cohen, p. 15.
28. Walton Family Foundation, "School Choice," available at http://www.waltonfamilyfoundation.org/educationreform/index.asp#2.
29. This is extensively detailed in Robert Greenwald's film *Wal-mart: The High Cost of Low Prices* (2005). An excellent source is the multi-article series on Wal-Mart by Liza Featherstone in *The Nation* available at http://www.thenation.com/search/?search=wal-mart.
30. Cohen, p. 16.
31. Bryan C. Hassel and Amy Way, "Chapter 7: Choosing to Fund School Choice" *With the Best of Intentions*, Cambridge: Harvard Education Press, 2005, p. 178.

32. Cohen, p. 6.

33. Thomas Pedroni, *Market Movements: African American Involvement in School Voucher Reform,* New York: Routledge 2007.

34. Cohen, p. 15.

35. As the economic crisis continued through the start of 2009, *New York Times* columnist Thomas L. Friedman who made his name with an unrelenting series of neoliberal books and articles began sounding surprisingly similar to the voices of the global justice movement with calls for a "zero growth economy" as the only alternative to the unsustainable and ecologically destructive neoliberal view of endless unchecked economic growth.

36. Kenneth J. Saltman, *The Edison Schools,* New York: Routledge 2005.

37. Cohen, p .14.

38. Cohen, p. 13.

39. Cohen, p. 21.

40. Cohen, p. 21.

41. Cohen, p. 19.

42. Cohen, p. 19.

43. Cohen, p. 22.

44. See Kenneth Saltman, "Chapter One: Silver Linings and Golden Opportunities" in *Capitalizing on Disaster: Taking and Breaking Public Schools,* Boulder: Paradigm, 2007.

45. Cohen, p. 24.

46. Cohen writes, "In researching the political contributions of 957 identifiable executive directors and board chairs of the think tanks and their funders, 368 had contributed to Republican candidates or causes during the 2000, 2002, 2004, 2006 and the beginning of the 2008 electoral cycle, while only 121 donated to Democratic candidates or vehicles. Donations to Republicans exceeded donations to Democrats by a ratio of more than 8 to 1, with $13,141,374 going to Republican Party candidates or vehicles compared to only $1,612,141 to the opposition Democrats." (pp. 24–25).

47. Hassel and Way, p. 178.

48. Cohen, p. 30.

49. Henry A. Giroux, *The Terror of Neoliberalism: The New Authoritarianism and the Eclipse of Democracy,* Boulder: Paradigm Publishers 2004, p. 106.

50. Hassel and Way, p. 183.

51. Hess, "Conclusion" *With the Best of Intentions.*

52. See, for example, Hassel and Way.

53. Alex Molnar, Gary Miron, Jessica Urschel, "Profiles of for Profit Educational Management Organizations: 10th Annual Report Commercialism in Education Resesarch Unit" Education in the Public Interest Center 2008, p. 10 available online at http://www.epicpolicy.org/publication/profiles-profit-education-management-organizations-2007–2008.

54. Molnar, et al., p. 10.

55. Molnar et al., p. 10

56. Personal attendance.

57. Available at http://newschools.org/work, http://www.chartergrowthfund.org

57. Press Release, "CPS Moves to Turn Around Six More Schools" 2009 available at http://www.cps.edu/News/Press_releases/2009/Pages/01_16_ 2009_ PR1.aspx.
58. Hassel and Way, p. 184.
59. Hassel and Way, p. 188.
60. Hassesl and Way, p. 189.
61. Hassel and Way, p. 190.
62. Hassel and Way, p. 192.

## CHAPTER 3

1. See Robert Arnove, *Philanthropy and Cultural Imperialism: The Foundations at Home and Abroad,* Bloomington: Indiana University Press, 1982 and William Watkins, *The White Architects of Black Education,* New York: Teachers College Press, 2001.
2. For discussions of neoliberal education, see Kenneth J. Saltman, *Collateral Damage: Corporatizing Public Schools—a Threat to Democracy,* Lanham, MD: Rowman & Littlefield 2000; Robin Truth Goodman and Kenneth J. Saltman, *Strange Love, Or How We Learn to Stop Worrying and Love the Market,* Lanham, MD: Rowman & Littlefield 2002; Henry A. Giroux, *The Terror of Neoliberalism: The New Authoritarianism and the Eclipse of Democracy,* Boulder: Paradigm Publishers 2004; Michael Apple, *Educating the Right Way,* New York: Routledge 2001.
3. See Kenneth J. Saltman, *Capitalizing on Disaster: Taking and Breaking Public Schools,* Boulder: Paradigm Publishers 2007 for a detailed account of the ways that neoliberal education has turned toward the pillage of public schooling and housing.
4. Stanley N. Katz, "Philanthropy's New Math" *The Chronicle of Higher Education,* 1F2 2007 supp., p. 2
5. Katz, p. 7.
6. Ira Silver, "Disentangling Class from Philanthropy: The Double-edged Sword of Alternative Giving" *Critical Sociology* 33 (2007) 538.
7. These are discussed in Joan Roelofs, "Foundations and Collaboration" *Critical Sociology* 33(2007) 479–504.
8. Joan Roelofs, p. 491.
9. This history is discussed in Stephen J. Gould, *Ontogeny and Phylogeny,* Cambridge, MA: Belknap Press of Harvard University Press, 1985; Gail Bederman, *Manliness and Civilization: A Cultural History of Gender and Race in the United States, 1880–1917,* Chicago: University of Chicago Press, 1996; Nancy Lesko, *Act Your Age!* New York: Routledge 2001; and Brown and Satman, *The Critical Middle School Reader,* New York: Routledge 2005; the last contains an exemplary excerpt from G. Stanley Hall's book *Adolescence.*
10. Georges Bataille, *The Accursed Share Vol. I.* New York: Zone Books, p. 126.
11. The PBS mini-series documentary *The Triumph of the Nerds: The Rise of Accidental Empires* (1996) provides a valuable history of the early commodification of the software and computer industry.

12. Richard Lee Colvin, p. 25. See Alex Molnar, *Giving Kids the Business,* New York: Westview 2001.
13. Colvin, 26.
14. Hess, "Introduction," p. 5.
15. Colvin, p. 32.
16. Hess, p. 11.
17. Frumkin, p. 8.
18. Frumkin, p. 8.
19. Frumkin, p. 8.
20. Hess, "Conclusion," p. 300.
21. Hassel and Way, 196.
22. Frederick M. Hess, "Conclusion" *With the Best of Intentions,* Cambridge: Harvard Education Press 2005, pp. 299–300.
23. For a defense of such a view of pluralism and reference to the vast literature from this perspective, see Peter Frumkin *Strategic Giving,* Chicago: University of Chicago Press 2007 and Joel Fleishman *The Foundation: A Great American Secret: How Private Wealth is Changing the World,* New York: Public Affairs 2007.

## CHAPTER 4

1. Christopher G. Robbins, *Expelling Hope: The Assault on Youth and the Militarization of Schooling,* Albany: SUNY Press 2008.
2. Henry A. Giroux, *Abandoned Generation: Democracy Beyond the Culture of Fear,* New York: Palgrave Macmillan 2004; Mike Males *Scapegoat Generation: America's War on Adolescents,* Monroe, ME: Common Courage Press 1996; Lawrence Grossberg, *Caught in the Crossfire,* Boulder: Paradigm Press 2005.
3. See David Harvey, *A Brief History of Neoliberalism,* Oxford: Oxford University Press 2005.
4. Online NewsHour, October 23, 2008, Transcript, "Greenspan admits 'flaw' to Congress, predicts more economic problems," available online at http://www.pbs.org/newshour.
5. BetsAnn Smith, "Deregulation and the New Leader Agenda: Outcomes and Lessons from Michigan," *Educational Administration Quarterly* V44 n1 2008, p. 51.
6. Smith, 51.
7. Frederick Hess and Andrew Kelly, "An Innovative Look, A Recalcitrant Reality: The Politics of Principle Preparation Reform" *Educational Policy* V19 n1, January and March 2005, p. 158.
8. Hess and Kelly, 158.
9. Hess and Kelly, 170.
10. Hess and Kelly, 171.
11. Hess and Kelly, 172.
12. Lesli A. Maxwell, "Challenging the Status Quo" *Education Week* June 21, 2006, V25 i41, 36–40.
13. Maxwell, 36–40.

14. Fenwick W. English, "The Unintended Consequences of a Standardized Knowledge Base in Advancing Educational Leadership Preparation" *Educational Administration Quarterly* V42 n3 August 2006, p. 461.

15. The attempts at longitudinal tracking of administrator "performance" and teacher "performance" and then isolating individual behaviors or methodological approaches has been funded by the Carnegie Corporation and the U.S. Department of Education under the Bush administration without success. Meanwhile, the Broad and Gates foundations are continuing to fund these kinds of positivist projects that share the same assumptions.

16. I detail neoliberal educational rebuilding in New Orleans in *Capitalizing on Disaster: Taking and Breaking Public Schools,* Boulder: Paradigm 2007.

17. AP, "Louisiana: Grants For Schools" *The New York Times*, December 14, 2007, p.33.

18. Matthew Pinzur, "Dade Selected as One of 5 Most-Improved Urban School Districts" *The Miami Herald*, April 6, 2006

19. Marianne Hurst, "California District Awarded Urban Education Prize" *Education Week* October 1, 2003.

20. Stanley Aronowitz, Stanley Aronowitz, *Against Schooling for An Education That Matters,* Boulder: Paradigm Publishers, 2008, p. 18.

21. Aronowitz, 23.

22. Henry Giroux, "Casino Capitalism" *truthout.org*; Kenneth Saltman, "Gambling with the Future of Public Education: risk, discipline, and the moralizing of educational politics in corporate media" Policy Futures in Education V5 n1 2007.

23. Lynn Olson, "State Chiefs, Businesses Forge $45 Million Data Venture" *Education Week* July 14, 2004, V23 i42, p. 16.

24. See, for example, the work of Donaldo Macedo and John Willinsky on this subject.

25. Richard Tomlinson and David Evans "CDO Boom Masks Subprime Losses, Abetted by S&P, Moody's, Fitch" *Bloomberg.com* available at http://www.bloomberg.com/apps/news?pid=newsarchive&sid=ajs7BqG4_X8I.

## CHAPTER 5

1. David F. Labaree, *The Trouble with Ed Schools*, New Haven, CT: Yale University Press, 1997.

2. See for example, Linda Darling-Hammond, "Teaching for America's Future: National Commissions and Vested Interests in an Almost Profession" *Educational Policy* 14(1), January and March 2000, 162–183.

3. Marilyn Cochran-Smith and Mary Kim Fries "Sticks, Stones, and Ideology" *Educational Researcher* 1, 3.

4. Ibid.

5. Pierre Bourdieu "The Forms of Capital" in J. Richardson (ed.) *Handbook of Theory and Research for the Sociology of Education*, New York: Greenwood, 1986, pp. 241–258.

6. Antonio Gramsci, *Selections from the Prison Notebooks* (ed.), Quintin Hoare and Geoffrey Nowell Smith New York: International Publishers, 1971. See "The Formation of the Intellectuals" pp. 3–13.

7. Lois Weiner, "A Lethal Threat to U.S. Teacher Education" *Journal of Teacher Education*, 58(4) September/October 2007, p. 283.
8. For an extensive criticism of the corporate hijacking of the small schools agenda and the Gates involvement in this, see Mike and Susan Klonsky, *Small Schools: The Smalls Schools Movement Meets the Ownership Society*, New York: Routledge, 2007.
9. Ibid.
10. The Oregonian, "Vickie Philips to Step Down as Portland Schools Superintendent" *The Oregonian*, April 25, 2007 available at http://blog.oregonlive.com/breakingnews/2007/04/vicki_phillips_to_step_down_as.html.
11. Ibid.
12. Ibid.
13. Ibid.
14. Ibid.
15. See Giroux's important *Teachers as Intellectuals*, Westport: Bergin & Garvey 1988 for an important distinction between the roles of different kinds of intellectual teachers. Giroux argues for teachers to be not only critical but also transformative.
16. "Obama defines his pragmatism for C-SPAN's Steve Scully" reprinted on Lynn Sweet: The Scoop from Washington in *Chicago Sun-Times* May 23, 2009 4:28 PM available at http://blogs.suntimes.com/sweet/2009/05/obama_defines_his_pragmatism_f.html.
17. Karen Symms Gallagher and Jerry D. Bailey, "The Politics of Teacher Education Reform: Strategic Philanthropy and Public Policy Making," *Educational Policy* 2000, 12.
18. Gallagher and Bailey, p. 21.
19. Gallagher and Bailey, p. 13.
20. Gallagher and Bailey, p. 14.
21. Gallagher and Bailey, p. 22.
22. Gallagher and Bailey, p. 22.

# CHAPTER 6

1. Gramsci's influence on Althusser is particularly evident by reading Gramsci's "The Modern Prince" in Antonio Gramsci, *Selections From The Prison Notebooks* (ed.), Quintin Hoare and Geoffrey Nowell Smith New York: International Publishers, 1971 with Althusser's *Machiavelli and Us*, New York: Verso, 1999.
2. See, for example, Michael Apple, *Ideology and Curriculum*, New York: Routledge, 1979 and Henry Giroux, *Teachers as Intellectuals*, Westport: Bergin & Garvey, 1988.
3. For examples of the recent Marxist educational thought that has largely eschewed the insights of much of critical theory, postrstructuralism, pragmatism, postcolonial theory, feminism, psychoanalysis, and critical race theory in favor of a return to class above all else, see the otherwise valuable and insightful work of Dave Hill, Glen Rikowski, Peter McLaren, Mike Cole, Rich Gibson, and Ramin Farahmandpur. Part of the problem here is that

rather than appropriating from these different traditions to strengthen and expand class analysis, some of the new old Marxism pits all other traditions of thought as the enemy of the One. This belies a dogmatic religiosity not to mention a failure to grasp the crucial criticisms of the logic of enlightenment. This perspective is a comprehensible over-reaction to some of the worst excesses of the postmodern trend in educational theory that resulted in the depoliticized insistance on localism, a myopic identity politics, the rejection of the category of class and political economic analysis, cultural relativism, the rejection of any narrative of emancipation or progress, the celebration of desire in ways that merely reinscribed consumerism to name a few. Nonetheless, these Marxist authors ought to embrace the selective appropriation of diverse theoretical tools. Part of the problem for McLaren, who has proven himself a brilliant theorist of culture, is that in order to talk about culture from this limited perspective, he is forced to smuggle the tools and select language of poststructural analysis back into the text while disavowing those very tools as a threat to a pure class analysis. Another tragic dimension to this educational trend is that it is out of step with actually existing radical left movements around the world including the Global Justice Movement and the movement toward socialism in Latin America. For example, as I finish edits to this book in Peru, Indians in the Peruvian Amazon resist the Peruvian government's attempts to steal their land to lease it to foreign nations for energy exploitation. People there link together the class, ethnic, racial, linguistic, and ecological struggles as a singular global movement. The average person understands these interrelated relationships and does not resort to the hierarchicizing of oppressions characteristic of the new old Marxism. As one taxi driver in Arequipa put it, the Indians are fighting the corporations not to make it their Amazon but because it is everyone's Amazon.

4. Jean Baudrillard, *The Mirror of Production*, Telos Press, 1975. In education, see Stanley Aronowitz and Henry Giroux *Education Still Under Siege*, Westport: Greenwood, 1989 for a brilliant and multfaceted dissection of the Marxist legacy.

5. See, for example, Aronowitz and Giroux, *Education Still Under Siege*, Westport: Bergin and Garvey, 1989.

6. Lawrence Grossberg, *Caught in the Crossfire*, Boulder: Paradigm Publishers 2005.

7. See Antonio Gramsci, "The Intellectuals" in *Selections from the Prison Notebooks*, edited and translated by Quintin Hoare and Geoffrey Nowell Smith, New York: International Publishers, p. 12.

8. Dave Hill, AERA Symposium "Tyrrany of Neoliberalism on Education" paper presentation titled "Analyzing and Resisting Capitalist Education: Class 'Race' and Contemporary Capitalism" p. 16.

9. Hill, ibid.

10. Chantal Mouffe *On the Political*, New York: Routledge, 2005, p. 20.

11. See, for example, John Chubb and Terry Moe's *Politics, Markets, and America's Schools*, Washington, DC: Brookings Institution Press (1990) for the classic neoliberal formulation of this position or Milton Friedman's *Capitalism and Freedom*, Chicago: University of Chicago Press, 1962.

12. Much of the literature on reproduction theory in education confronts this. See for example, Bowles and Gintiss's classic *Schooling in Capitalist America* the forms of capital addressed by Pierre Bourdieu as well as the more recent literature on neoliberalism in education by authors such as Giroux, Hursh, Goodman, Buras, Apple, and my books *Collateral Damage, The Edison Schools*, and *Capitalizing on Disaster.*

13. I refer here to the sense of praxis suggested by Paulo Freire in which relection on experience leads to collective action which in turn needs to be theorized toward the end of transforming oppressive social forces.

14. Marcel Mauss, *The Gift: The Form and Reason for Exchange in Archaic Societies*, New York: Norton, 1990. (Originally published in French in 1922 and in English in 1954)

15. While I am familiar with the accusations of Mauss's primitivism for celebrating the so-called archaic societies and also for missing what Marx pointed out about the radically beneficial creative destruction of capitalism in that, for example, it challenges "the idiocy of rural life" that is, traditions that ought to go, I opt here to sidestep these issues in order to appropriate what might be gleaned for getting around economism left and right. It would be extremely valuable to elsewhere take up the extent to which neoliberal capitalism brings back to some extent the "total system in services" by radically expanding markets to all aspects of social life. The crucial issue is how the obligations that get produced in neoliberal capitalism are radically individualized rather than collective such that collective problems can only be addressed individually. This is getting "terminal" in the sense that the survival of the planet ecologically hinges on collective solutions to collective problems. The neoliberal model of unlimited growth of profits along with individualization of collective problems assures collective global destruction.

16. For an excellent discussion of Mauss and his critics and theories of the gift generally, see David Graeber, *Toward and Anthropological Theory of Value: The False Coin of Our Own Dreams*, New York: Palgrave, 2001.

17. Donaldo Macedo's *Literacies of Power: What Americans are Not Allowed to Know* New, York: Westview Press (1993) elaborates on the process of "stupidification" in an enduring and valuable way.

18. Mauss' perspective does have some limitations for thinking education beyond economism. First, in the courses of making the entire social world economic, as Marshal Sahlins highlights, Mauss lacks an adequate theory of political society. See Marshal Sahlins's *Stone Age Economics*, New York: Transaction Publishers, 1972. The relevant criticisms are reprinted in Alan Schrift (ed.), *The Logic of the Gift*, New York: Routledge (1992), which is one of the best collections of literature on the gift. For Sahlins, there is an inability to adequately theorize civil society. Second, a point that Levi-Strauss first raises, Mauss's understanding of what produces the obligation to reciprocate should not be thought in terms of the culture in question but should be deduced from the system of rules found transculturally. Hence, as Levi-Strauss suggests, Mauss put forward two explanations for what produces the relation of obligation. One comes from the Maori Hau, the spirit of the gift. It involves the drive to give oneself when one gives. The other explanation is the rules of the system of exchange are generated from the system of exchange.

## CONCLUSION

1. Pierre Bourdieu, "Marginalia—Some Additional Notes on the Gift" in *The Logic of the Gift*, (ed.) Alan Shrift, New York: Routledge, 1992, p. 240.

## CODA

1. Cited in Alfie Kohn, "The Real Threat to American Schools," Tikkun (March–April 2001), p. 25. For an interesting commentary on Obama and his possible pick to head the education department and the struggle over school reform, see Alfie Kohn, "Beware School 'Reformers,'" The Nation (December 29, 2008). Available online at www.thenation.com/doc/20081229/kohn/print.
2. This term comes from David Garland, *The Culture of Control: Crime and Social Order in Contemporary Society*, Chicago: University of Chicago Press, 2002.
3. For a brilliant analysis of the "governing through crime" complex, see Jonathan Simon, *Governing Through Crime: How the War on Crime Transformed American Democracy and Created a Culture of Fear*, New York: Oxford University Press, 2007.
4. Advancement Project in partnership with Padres and Jovenes Unidos, Southwest Youth Collaborative, *Education on Lockdown: The Schoolhouse to Jailhouse Track*, New York: Children & Family Justice Center of Northwestern University School of Law, March 24, 2005, p. 31. On the broader issue of the effect of racialized zero-tolerance policies on public education, see Christopher G. Robbins, *Expelling Hope: The Assault on Youth and the Militarization of Schooling*, Albany: SUNY Press, 2008. See also, Henry A. Giroux, *The Abandoned Generation*, New York: Palgrave, 2004.
5. David Hursh and Pauline Lipman, "Chapter 8: Renaissance 2010: The Reassertion of Ruling-Class Power through Neoliberal Policies in Chicago" in David Hursh, *High-Stakes Testing and the Decline of Teaching and Learning*, Lanham, MD: Rowman & Littlefield, 2008.
6. See online at www.atkearney.com
7. Creating a New Market of Public Education: The Renaissance Schools Fund 2008 Progress Report, The Renaissance Schools Fund at www.rsfchicago.org.
8. Kenneth J. Saltman, Chapter 3: *Renaissance 2010 and No Child Left Behind Capitalizing on Disaster: Taking and Breaking Public Schools*" (Boulder: Paradigm Publishers, 2007).
9. Sarah Karp and Joyn Myers, "Duncan's Track Record," Catalyst Chicago, December 15, 2008. Available online at www.catalyst-chicago.org/news/index.php?item=2514&cat=5&tr=y&auid=4336549.
10. See Chicago Public Schools Office of New Schools 2006/2007 Charter School Performance Report Executive Summary.
11. See Dorothy Shipps, *School Reform, Corporate Style: Chicago 1880–2000*, Lawrence: University of Kansas Press, 2006.
12. See, for example, Summary Report, "America's Cradle to Prison Pipeline," Children's Defense Fund. Available online at www.childrensdefense.org/

site/DocServer/CPP_report_2007_summary.pdf?docID=6001; also see, Elora Mukherjee, "Criminalizing the Classroom: The Over-Policing of New York City Schools," New York: American Civil Liberties Union and New York Civil Liberties, March 2008, pp. 1–36.

13. Donna Gaines, "How Schools Teach Our Kids to Hate," Newsday, Sunday, April 25, 1999, p. B5.

14. As has been widely, reported, the prison industry has become big business with many states spending more on prison construction than on university construction. Jennifer Warren, "One in 100: Behind Bars in America 2008," (Washington, DC: The PEW Center on the States, 2007). Available online at www.pewcenteronthestates.org/news_room_detail.aspx?id=35912.

# Index